F. Hall

C.Ed, H.N.C., C.G.F.T.C., R.P., M.N.A.P.T.
City and Guilds Silver medallist in Plumbing

Design calculations for plumbing and heating engineers

Longman
Scientific &
Technical

Longman Scientific & Technical,
Longman Group UK Limited,
Longman House, Burnt Mill, Harlow,
Essex CM20 2JE, England
and Associated Companies throughout the world.

© Longman Group Limited 1982
This edition © Longman Group UK Limited 1988

First published 1982
Second edition 1988

British Library Cataloguing in Publication Data

Hall, F.
    Design calculations for plumbing and
    heating engineers. — 2nd ed.
    1. Plumbing — Mathematics — Problems,
    exercises, etc.   2. Heating — Mathematics
    — Problems, exercises, etc.
    I. Title   II. Hall, F.
    696'.1'0151   TH7226

    ISBN 0-582-01867-6

Produced by Longman Singapore Publishers (Pte) Ltd.
Printed in Singapore.

Contents

Preface

During his studies at the technical college, the plumbing and heating student carries out various calculations related to his craft. These calculations include force, pressure, pipe-sizing, heat losses, heating costs, and radiator and boiler sizing, etc. All these subjects have been dealt with in the text and in order to show how these are carried out, practical examples have been given.

The text covers closely the calculations required by the City and Guilds of London Institute syllabus in both plumbing and heating. Whenever necessary, diagrams and tables have been included to clarify the examples given in the text. Questions with answers have been included at the end of each chapter and it will be to the student's advantage to test his knowledge by carrying them out

Although the book is written specifically for students, the practicing plumbing or heating craftsman should also find it useful. SI units are used throughout the book and a description of the units has been included in the appendix. Surface areas, volumes and perimeters have also been included in the appendix. Chapter 7, Heating, covers the essentials of the Building Regulations 1985 dealing with conservation of fuel and power.

I should like to thank C. R. Bassett for his helpful constructive criticisms, and a special thank you to my wife for her patience and understanding during the preparation of the book and for typing the manuscript.

F. Hall 1988

Acknowledgements

We are grateful to the following for permission to reproduce copyright material:

Chartered Institution of Building Services for our Tables 8.1, 8.8, 8.10 and 8.11 from *CIBS Guide*; The Wednesbury Tube Company for our Tables 8.5 and 8.6.

Chapter 1

Speed, velocity and acceleration

Speed, velocity and acceleration

Speed is defined as the rate of change of distance moved with time. Velocity is defined as the rate of change of distance moved with time in a specified direction. Speed is a scalar quantity whereas velocity is a vector.

When travelling by car it is speed that can be read on the speedometer because the direction is unimportant. A speed of 50 km/h is known as a scalar quantity while a velocity of 50 km/h due south is a vector. Velocity can therefore be represented by a straight line drawn to scale in a given direction.

Average speed

Unless travelling on a motorway, a motorist cannot usually maintain a constant speed for any length of time. When deciding the time to allow for a particular journey, the motorist must have some idea of the average speed at which it is possible to travel.

Example 1.1. *Find the average speed during a journey of 276 km which takes 6 hours.*

$$\text{average speed} = \frac{\text{distance}}{\text{time}} = \frac{276}{6} = 46 \text{ km/h}$$

Uniform velocity

A body is said to move with uniform velocity if its rate of change of distance with time in a specified direction is constant.

In services the distance and time are expressed in metres and seconds respectively.

$$\text{velocity} = \frac{\text{distance}}{\text{time}} = \frac{\text{metres}}{\text{seconds}} = \text{m/s}$$

Example 1.2. *A travelator (moving pavement) moves at a velocity of 0.6 m/s. Calculate the distance a person would move on the travelator in 30 seconds.*

$$\text{velocity} = \frac{\text{distance}}{\text{time}}$$

$$\text{distance} = \text{time} \times \text{velocity}$$
$$= 30 \times 0.6$$
$$= 18 \text{ m}$$

Acceleration

Acceleration is a natural progression from velocity, and is defined as the rate of change of velocity with time.

$$\text{acceleration} = \frac{\text{velocity}}{\text{time}} = \text{m/s}^2$$

Example 1.3. *A lift car accelerates uniformly from rest to 2 m/s in 4 seconds. Calculate the acceleration and the distance travelled by the car.*

$$\text{acceleration} = \frac{\text{change in velocity}}{\text{time}}$$

$$= \frac{2 \text{ m/s}}{4 \text{ s}}$$

$$= 0.5 \text{ m/s}^2$$

In order to calculate the distance travelled by the car, the average velocity is required.

$$\frac{\text{average velocity for}}{\text{uniform acceleration}} = \frac{\text{initial velocity + final velocity}}{2}$$

$$\text{average velocity} = \frac{0 + 2}{2}$$

$$= 1 \text{ m/s}$$
$$\text{distance travelled} = 1 \times 4$$
$$= 4 \text{ m}$$

Example 1.4. *A lift car starts from rest and is accelerated uniformly at the rate of 0.5 m/s² for 5 s. It then maintains a constant velocity for 10 s. The car is then uniformly retarded at the rate of 0.5 m/s² for 5 s. Find the maximum velocity of the car and the total distance it travels.*

Method 1. By calculation
u = initial velocity in m/s
a = acceleration in m/s^2

v = final or maximum velocity in m/s
t = time for which acceleration continues in seconds
s = distance travelled in m

To find the maximum velocity

u = 0 m/s
a = 0.5 m/s^2
t = 5 s
v = 0 + 0.5 × 5
   = 2.5 m/s   which is the maximum velocity

First stage.   Distance travelled is the average velocity m/s multiplied by time in seconds.

$$s = \frac{2.5}{2} \times 5$$

s = 6.25 m

Second stage.   distance moved = velocity × time
                            = 2.5 × 10
                            = 25 m

Third stage.   The distance travelled in the third stage will be the same as that in the first stage.

total distance travelled = 6.25 + 25 + 6.25
                        = 37.5 m

Method 2. Graphical solution

Draw the velocity–time graph (Fig. 1.1).

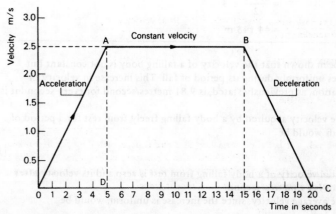

Fig.1.1   Velocity-time graph for a lift car

3

The area under the graph gives the distance travelled by the car.

$$\text{area OABC} = \frac{1}{2} (AB + OC) \times AD$$

$$= \frac{1}{2} (10 + 20) \times 2.5$$

$$= 37.5 \text{ m}$$

Gravitational acceleration

A well-known rate of acceleration is that due to gravity. If a body is falling freely near to the surface of the earth it has an acceleration due to gravity of 9.81 m/s^2, which is applicable to most geographical locations.

The velocity of a falling body does not depend upon mass but upon its nearness to the surface of the earth. In the study of dynamics in the sixteenth century, the Italian, Galileo, released simultaneously three iron balls of different masses from the top of the Leaning Tower of Pisa, and all three balls reached the ground at the same time. Before this time it was thought that the velocity of a falling body depended upon mass.

Example 1.5. *An iron ball is dropped from the top of a building and takes 3 seconds to reach the ground.*

(a) What is its velocity when it hits the ground?
(b) What is the height of the building?

terminal velocity = acceleration due to gravity $\times$ time
$$= 9.81 \times 3$$
$$= 29.43 \text{ m/s}$$

height of building = average velocity $\times$ time
(distance fallen)

$$= \frac{29.43}{2} \times 3$$

$$= 44.145 \text{ m}$$

It has been shown that the velocity of a falling body is not constant but increases uniformly during its period of fall. This increase in velocity or acceleration, as previously stated, is 9.81 metres/second for every second it is falling.

The velocity acquired by a body falling freely from rest for a period of t seconds would be

$$v = g\,t \qquad\qquad\qquad [1.1]$$

The initial velocity of a body falling from rest is zero and its velocity after t seconds is $g\,t$ metres/second.

The average velocity, since the increase is uniform, would be

$$\frac{\text{average velocity}}{\text{m/s}} = \frac{0 + g\,t}{2}$$

$$= \frac{g\,t}{2}$$

The distance fallen will be average velocity × time of fall

$$\text{distance fallen} = \frac{g\,t}{2} \times t$$

$$= \frac{g\,t^2}{2}$$

Substituting from equation [1.1]

$$t = \frac{v}{g}$$

$$b = \frac{g}{2} \times \frac{v}{g} \times \frac{v}{g}$$

$$= \frac{v^2}{2\,g}$$

Kinetic or velocity energy

The expression $b = v^2/2\,g$ gives the kinetic or velocity energy used in hydraulic calculations, and by transposing the formula, the theoretical velocity of water flowing through an orifice or a pipe may be found.

$$v = \sqrt{2\,g\,b} \tag{1.2}$$

Vena contracta

Jet of water

Fig.1.2 Flow of water through an orifice

5

where v = theoretical velocity of flow in m/s
$\quad\quad g$ = 9.81 m/s^2
$\quad\quad h$ = head of water in metres above the centre of the orifice or pipe

Due to the effect of friction, the theoretical velocity is not fully reached. When a jet of water is discharging freely through an orifice, the particles are trying to come together in all directions (see Fig. 1.2). This contraction is known as 'vena contracta' and is found by experiment to be about 0.64 of the orifice area. This is also termed the coefficient of contraction.

The effect of friction, known as the coefficient of velocity, is about 0.97 and the product of the coefficient of contraction and the coefficient of friction is known as the coefficient of discharge.

the coefficient of discharge $Cd = Cc \times Cv$

where Cc = coefficient of contraction
$\quad\quad Cv$ = coefficient of velocity
$\quad\quad \therefore Cd = 0.64 \times 0.97 = 0.62$ (approx.)

The discharge through a pipe or orifice may be found from the formula

$$Q = VA$$

where Q = rate of discharge in m^3/s
$\quad\quad V$ = velocity of flow in m/s
$\quad\quad A$ = area of pipe or orifice in m^2

Substituting from equation [1.2]

$$Q = Cd \times A \times \sqrt{2\,g\,h}$$

Example 1.6. *Calculate the discharge in litres per second through a 100 mm diameter orifice at the base of a tank when there is a constant head of water above the centre of the orifice of 1.5 m.*

$\quad\quad Q = Cd \times A \times \sqrt{2\,g\,h}$
$\quad\quad\quad = 0.62 \times 3.142 \times 0.05 \times 0.05 \times \sqrt{2 \times 9.81 \times 1.5}$
$\quad\quad\quad = 0.026\ 4$ m^3/s
discharge = 26.4 litre/s

If a short length of pipe of the same diameter of the orifice is fixed to the tank, a different coefficient of discharge will result, depending upon the type of connection.

Figure 1.3 shows three different types of connections to the tank, and it will be seen that a bell-shaped connection will give the best discharge.

Example 1.7. *Calculate the time required to discharge 500 litres of water from a tank by means of a 50 mm diameter short pipe having a bell-shaped connection to the tank when there is a constant head of water above the centre of the pipe of 1.5 m.*

$Q = Cd \times A \times \sqrt{2\,g\,h}$
$\quad\quad = 0.97 \times 3.142 \times 0.025 \times 0.025 \sqrt{2 \times 9.81 \times 1.5}$
$\quad\quad = 0.001\ 905 \times 5.425$
$\quad\quad = 0.010\ 335$ m^3/s
$\quad\quad\quad$ discharge = 10.335 litre/s

6

Fig.1.3 Pipe connections to tanks and cisterns

Bell-shaped *Cd* = 0.97

Square *Cd* = 0.9

Re-entrant *Cd* = 0.5

Water level

Water level

Water level

Vena contracta

Vena contracta

$$\text{time taken} = \frac{500}{10.335}$$

$$= 48.38 \text{ s}$$

The value of $Q = \pi R^2 \sqrt{2 g h}$ will be constant for each case but the coefficient of discharge will of course vary.

$Q = 3.142 \times 0.05 \times 0.05 \times \sqrt{2 \times 9.81 \times 1.5}$
$= 0.007\ 855 \times \sqrt{29.43}$
$= 0.042\ 61 \text{ m}^3/\text{s}$ (theoretical)

Q (re-entrant outlet) $= 0.042\ 61 \times 0.5$
$\qquad\qquad\qquad\quad = 0.0213 \text{ m}^3/\text{s}$
$\qquad\qquad\qquad\quad = 21.30 \text{ litre/s}$

Q (square outlet) $= 0.042\ 61 \times 0.9$
$\qquad\qquad\qquad = 0.038\ 35 \text{ m}^3/\text{s}$
$\qquad\qquad\qquad = 38.35 \text{ litre/s}$

Q (bell-shaped outlet) $= 0.042\ 61 \times 0.97$
$\qquad\qquad\qquad\qquad = 0.041\ 33 \text{ m}^3/\text{s}$
$\qquad\qquad\qquad\qquad = 41.33 \text{ litre/s}$

Questions

1. A lift car starts from rest and is accelerated uniformly at the rate of 0.2 m/s^2 for 3 seconds. It then maintains a constant velocity for 8 seconds. The car then decelerates uniformly at the rate of 0.2 m/s^2 for 3 seconds. Find the maximum velocity of the car and the total distance it travels.

Answers: 0.6 m/s; 6.6 m

2. Draw a velocity–time graph for the movement of the lift car in question 1.

3. Define: (a) speed; (b) velocity; (c) acceleration; (d) gravitational acceleration.

4. An escalator moves at a velocity of 0.6 m/s. If the steps in use make up a distance of 8 m, calculate the time it would take a person to move up one floor.

Answer: 13.33 seconds

5. Calculate the velocity of water in a pipe of cross-sectional area of 0.02 m^2 which delivers 115 litres of water per second.

Answer: 5.75 m/s

6. Derive the formula for kinetic or velocity energy of a falling body.

7. Define: (a) vena contracta; (b) coefficient of discharge; (c) coefficient of velocity.

8. Compare the discharge of water through an orifice and the following short pipe connections to a tank: (a) re-entrant; (b) square; (c) bell-shaped.

8

9. Calculate the discharge in litres per second through a 50 mm diameter orifice at the side of a tank when there is a constant head of water above the centre of the orifice of 2 m.

Answer: 24.11 litre/s

10. Calculate the discharge in litres per second and the time required to discharge 300 000 litres of water from a swimming pool by means of a 76 mm diameter short pipe having a bell-shaped connection when there is a constant head of water above the centre of the pipe of 2 m.

Answers: 27.56 litre/s; 3 hours (approx.)

Chapter 2

Force and pressure

Force

The SI provides a coherent system of units in which the unit of force, the newton, can be used for calculations involving pressure, stress, work and power.

A force of 1 newton (N) is defined as that force which gives a body having a mass of 1 kilogram an acceleration of 1 metre per second squared (see Fig. 2.1).

Fig.2.1   Newton of force

force = mass × acceleration
∴ 1 newton = 1 kg × 1 m/s^2

It was stated in Chapter 1 that the acceleration due to gravity of a body falling freely was 9.81 m/s^2. Since the newton gives a 1 kilogram mass an acceleration of 1 metre per second squared, a kilogram mass falling freely will have acting upon it a force of 9.81 newton. This force is known as kilogram force and is used in calculations involving the pressure of water and other liquids.

Figure 2.2 shows a tank containing 1 cubic metre of water, which has a mass of approximately 1000 kg.

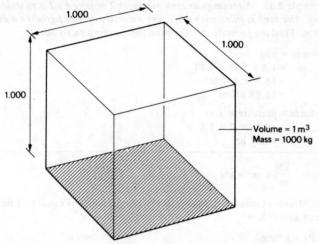

Pressure on the base due to 1 m head of water = 1000 x 9.81 N/m² = 9810 pascals or 9.81 kN/m² or 9.81 kPa

Fig.2.2 Pressure due to 1 m³ of water

The force acting over the base of 1 square metre due to the head of water of 1 metre will be:

force = mass × acceleration
= 1000 × 9.81
= 9810 N

Pressure of water

Since this force is acting over 1 square metre, the pressure on the base will be 9810 N/m², i.e.

$$\text{pressure} = \frac{N}{m^2} = N/m^2$$

Since the pressure is proportional to head of water, the pressure for all calculations dealing with water will be:

$$N/m^2 = 1000 \times 9.81 \times \text{head of water in metres}$$

The newton per square metre is a small unit for water pressure and it is usually more convenient when dealing with water to use kilonewton per square metre.

$$kN/m^2 = 9.81 \times \text{head of water in metres}$$

For liquids other than water the pressure may also be found because:

$$\begin{array}{c}\text{pressure due} \\ \text{to a liquid}\end{array} = \begin{array}{c}\text{head or} \\ \text{height}\end{array} \times \text{density} \times \begin{array}{c}\text{acceleration} \\ \text{due to gravity}\end{array}$$

or $\begin{array}{c}\text{pressure due} \\ \text{to a liquid}\end{array} = h\rho g$

11

Example 2.1. *A rectangular tank measures 2 m long by 1.5 m wide by 2 m deep. The tank is filled with oil having a density of 900 kg/m³ to a depth of 1.8 m. Find the pressure and the total force acting on its base.*

pressure = $h\rho g$

$ = 1.8 \times 900 \times 9.81$

$ = 15\ 892.2\ N/m^2$

$ = 15.89\ kN/m^2$

force = pressure × area

$ = 15.89 \times 2 \times 1.5$

$ = 47.67\ kN$

Note: $\dfrac{kN}{m^2} \times m^2 = kN$

The SI unit of pressure is known as the pascal, which is equal to 1 newton per metre squared, or

$1\ Pa = 1\ N/m^2$

$1\ kPa = 1\ kN/m^2$

Example 2.2. *Calculate the pressure of water acting upon a valve when the head of water above the valve is 15 m.*

pressure in Pa = head in metres × 1000 kg × 9.81

pressure in kPa = head in metres × 9.81

$ = 15 \times 9.81$

$ pressure = 147.15\ kPa$

When the pressure is known, it is sometimes necessary to find the head of water. Head is defined as the vertical distance between the free surface of the liquid and the point under consideration.

The pressure acting is proportional to head, and volume of liquid does not affect pressure. The pressure at the base of a tank holding 3 m³ of water when the head is 1.5 m is exactly the same as the pressure on the base of the tank holding 150 m³ with the same head of 1.5 m. The force acting on the base of the large tank, however, is much greater.

Example 2.3. *Calculate the pressure on the base of a tank measuring 4 m by 2 m by 2 m deep when the head of water is 1.5 m; also calculate the force acting on the base.*

pressure = 9.81×1.5

$ = 14.715\ kPa\ or\ kN/m^3$

$ = \dfrac{force}{area}$

force = pressure × area

$ = 14.715 \times 1.5$

$ = 22.0725\ kN$

$ = 22\ kN\ (approx.)$

12

If the area is less than 1 metre squared, the pressure and force can still be found, as illustrated in example 2.4.

Example 2.4. *Calculate the pressure acting on a valve and the force acting on its horizontal seating of 50 mm radius when the head of water above the valve is 15 m.*

pressure acting
on valve $= 9.81 \times 15$
             $= 147.15$ kPa

force acting
on valve $= 147.15 \times \pi R^2$
             $= 147.15 \times 3.142 \times 0.05 \times 0.05$
             $= 1.156$ kN

Example 2.5. *A pressure gauge on a boiler fed from a cistern shows a reading of 350 kPa. Calculate the head of water above the gauge.*

pressure = head in metres $\times$ 9.81

$$head = \frac{pressure}{9.81}$$

$$= \frac{350}{9.81}$$

$$= 35.68 \text{ m}$$

Absolute pressure

When dealing with problems in thermodynamics, absolute pressure is used.

$$\frac{absolute}{pressure} = \frac{gauge}{pressure} + \frac{atmospheric}{pressure}$$

Atmospheric pressure at sea-level will support a column of mercury 760 mm high (see Fig. 2.3). Since mercury is approximately 13.6 times heavier than water, atmospheric pressure will support a column of water:

$760 \times 13.6 = 10\ 336$ mm high
$= 10.33$ m
atmospheric pressure $= 10.33 \times 9.81 = 101.33$ kN/m^2
$= 101$ kPa (approx.)

Example 2.6. *Calculate the absolute pressure acting on top of a boiler when the head of water above the top is 20 m.*

gauge pressure $= 9.81 \times 20$
$= 196.2$ kPa
absolute pressure $= 196.2 + 101$
$= 297.2$ kPa

For ordinary pressure calculations, gauge pressure is used because atmospheric pressure is acting in opposite directions equally at the point under consideration and is therefore cancelled out. In example 2.6 the atmospheric pressure acting

Vacuum

760 mm

Atmospheric pressure
101 kPa approximately

Mercury

Fig.2.3 The mercury barometer

upon the surface of the water in the feed cistern is cancelled out by the atmospheric pressure acting outside on top of the boiler (see Fig. 2.4).

Example 2.7. *A water-filled manometer is connected to a gas pipe in order to measure the pressure of gas in the pipe. If the reading shows 200 mm head of water (water gauge), calculate the gas pressure in pascals and kilopascals.*

$$200 \text{ mm} = 0.2 \text{ m head}$$
$$\text{pressure in pascals} = \text{head in metres} \times 1000 \times 9.81$$
$$= 0.2 \times 1000 \times 9.81$$
$$= 1962 \text{ Pa}$$
$$= 1.962 \text{ kPa}$$

Gas pressure is sometimes measured in millibars or bars; these are not SI units but their values will be shown.

$$1 \text{ bar} = 10^5 \text{ N/m}^2 \text{ or } 100 \text{ kN/m}^2$$
$$1 \text{ mbar} = 100 \text{ N/m}^2 \text{ or } 0.1 \text{ kN/m}^2$$

$$\text{pressure in mbars} = \frac{1962}{100}$$
$$= 19.62$$

14

$$\text{pressure in bars} = \frac{1.962}{100}$$

$$= 0.019\ 62 \text{ bars}$$

Note: 1 bar is approximately equal to atmospheric pressure.

Feed cistern

Atmospheric pressure = 101 kPa (approx)

Feed pipe

Head of water
= 20.000

Gauge would show only
gauge pressure

Absolute pressure =
water pressure plus
atmospheric pressure

Boiler

Fig.2.4 Absolute pressure on top of boiler

The hydraulic press or jack (see Fig. 2.5)

The hydraulic press is a machine which consists of two different size pistons,
enabling it to lift weights by the application of a much smaller force. The weight
that can be lifted depends upon the ratio of the areas of the two pistons. When
the pistons are at the same height, the pressure of water acting upon both
pistons is the same. The force acting on each piston is equal to the pressure
multiplied by the area.

Example 2.8. *A hydraulic press has pistons having areas of 1000 mm² and
10 000 mm² respectively. What force is acting on the larger piston if a force of
10 kN is applied on the smaller one?*

Fig.2.5 The hydraulic press

ratio of pistons' areas $= \dfrac{10\ 000}{1000}$

$$= 1 : 10$$

Since water is virtually incompressible, the force acting upon the larger piston, neglecting friction, will be ten times the force acting on the smaller piston.

force acting to
lift the load   = 10 kN × 10
force on larger
piston   = 100 kN

Alternatively, the problem may be solved by finding the pressure acting upon the smaller piston.

pressure $= \dfrac{\text{force}}{\text{area}} = \dfrac{N}{mm^2}$

$$= \dfrac{10\ 000}{1000}$$

$$= 10 \text{ N/mm}^2$$

Since this pressure is the same on the larger piston, the lifting force may be found.

force = pressure × area of larger piston
    = N/mm² × mm²
    = 10 × 10 000 N
force on larger piston = 100 kN

Oil-hydraulic lift

The oil-hydraulic lift uses the principle of the hydraulic press. Oil at high pressure is pumped through a pipe to a cylinder containing a piston. The force acting upon the piston is equal to the pressure multiplied by the area of the piston. This force lifts the car plus its load.

Force on vertical surfaces (centre of pressure)

So far, formulae has been derived relating the head in metres with pressure in kilopascals and force on horizontal surfaces. Since water presses equally in all directions, consideration can now be given to surfaces which are vertical, such as the sides of cylinders, tanks, cisterns, access covers and sluice plates. The total force on any surface immersed in water is equal to the area of the surface multiplied by the pressure acting at the centre of the area (or the centroid) of the surface.

Example 2.9. (see Fig. 2.6): *A vertical rectangular plate is 1 m high by 1.200 m wide is acting as a sluice. The centre of the plate is 3 m below the surface of the water. Find the total force in kN acting on the plate.*

Note: The single resultant force acts through the centre of pressure

Fig.2.6 Force acting on a submersed plate

$$\text{depth or head to centre of area} = \bar{y} = 3 \text{ m}$$
$$\text{area of sluice plate} = 1 \times 1.2 = 1.200 \text{ m}^2$$
$$\text{pressure at centre of plate} = \bar{y} \times 9.81$$
$$= 3 \times 9.81$$
$$= 29.43 \text{ kPa}$$

$$\text{pressure kPa} = \frac{\text{force kN}}{\text{area m}^2}$$

17

$$\therefore \text{force} = \text{pressure kPa} \times \text{area m}^2$$
$$= 29.43 \times 1.200$$
$$= 35.316 \text{ kN}$$

Liquid contained in a vessel will exert a force at right angles to the surface of the vessel. The force will act upon each small element of area and the total force will be equal to the sum of these forces acting upon each small area.

total force acting = pressure at the centroid × total area.

Example 2.10 (see Fig. 2.7). *A water test is applied to a 100 mm diameter drain. Calculate the force acting upon the stopper when the head of water above the centre of the stopper is 4.05 m.*

Fig. 2.7 Force acting during drain testing

pressure at centroid = $\bar{y}$ (m) × 9.81
$$= 4.05 \times 9.81$$
$$= 39.73 \text{ kPa}$$

force acting upon
stopper = pressure (kPa) × area (m²)
$$= 39.73 \times 3.142 \times 0.05 \times 0.05$$
$$= 0.312 \text{ kN}$$
$$= 0.3 \text{ kN (approx.)}$$

Example 2.11 (see Fig. 2.8). *Assuming a static head of water, calculate: (a) the static pressure in kPa at A, B and C; (b) the force or thrust acting upon the circular manlid.*

pressure at A = head (m) × 9.81
$$= 4 \times 9.81$$
$$= 39.24 \text{ kPa}$$

pressure at B = head (m) × 9.81
$$= 7 \times 9.81$$
$$= 68.67 \text{ kPa}$$

pressure at C = $\bar{y}$ (m) × 9.81
$$= 10 \times 9.81$$
$$= 98.1 \text{ kPa}$$

Cold feed

4.000

(A)

3.000

Secondary return

(B)

Secondary flow

3.000

Calorifier

Primary flow

Manlid

Primary return

(C)

300 mm

Note: represents a
tap or valve

Fig.2.8   Pressures in a hot-water system

force acting
upon circular
manlid   = pressure (kPa) × area (m^2)
               = 98.1 × 3.142 × 0.15 × 0.15
               = 6.935 kN

Example 2.12 (see Fig. 2.8). *Calculate the force acting upon the total horizontal surface of the hot-water calorifier if its length is 2 m and internal diameter is 1.5 m.*

internal horizontal
surface   = πD × length
               = 3.142 × 1.5 × 2
               = 9.426 m^2

pressure at centre
of calorifier   = $\bar{y}$ (m) × 9.81
               = 10 × 9.81
               = 98.1 kPa

        force = pressure × area
               = 98.1 × 9.426
               = 924.6906 kN
               = 924.7 kN (approx.)   (924 700 N)

This is a tremendous force and a cylindrical vessel will provide the greatest strength to resist it. A rectangular vessel would have to be constructed with a much greater thickness of metal.

Example 2.13 (see Fig. 2.9). *A rectangular cistern measures 3 m long by 2 m wide by 1.5 m deep and contains water to the depth of 1.4 m. Find by means of a pressure diagram the resultant force acting on one vertical end.*

Pressure increases proportionately with the depth. The pressure as shown in Fig. 2.9 is represented by a triangle increasing from zero at the top to maximum at the base.

pressure at the base = 1.4 × 9.81
                         = 13.734 kPa

Since the distribution of pressure is represented by a triangle, the resultant force acts through the centre of the area of the triangle, that is one-third of the height above the base.

       The resultant force on the end may be found from the area of the pressure diagram.

area of triangle = $\frac{1}{2}$ AB × AC

resultant force
per (m) run   = $\frac{1}{2}$ × 1.4 × 13.734 kPa

 force on end = $\frac{1}{2}$ × 1.4 × 13.734 × 2

resultant force = 19.2276 kN

Fig.2.9 Pressure diagram showing the variation of pressure with depth

Check: The resultant force may also be found as follows:

resultant force = pressure at $\frac{1}{2}$ depth × area of end in contact with water

$$= 0.7 \times 9.81 \times 1.4 \times 2$$
$$= 19.2276 \text{ kN}$$
$$= 19.23 \text{ kN (approx.)}$$

Overturning moment (see Fig. 2.9)

If the resultant force acting on the end of the cistern and the position where this force acts are known, the overturning moment which tends to overturn the end may be found.

overturning moment = resultant force × distance from base

$$= 19.2276 \times \frac{1.4}{3}$$

$$= 8.972\,88 \text{ kNm}$$
$$= 9 \text{ kNm (approx.)}$$

21

In order to overcome this overturning moment, struts may be fixed across the top of the cistern. In order to resist the resultant force acting one-third of the head of water above the base, the metal should be made of sufficient strength and thickness, otherwise struts will also be required at this position.

Pressures on ballvalves

Ballvalves are classified according to the pressure that they are capable of closing against.

1. High-pressure valves should be capable of closing against a pressure of 1379 kPa.
2. Medium-pressure valves should be capable of closing against a pressure of 689.5 kPa.
3. Low-pressure valves should be capable of closing against a pressure of 276 kPa.

Every valve must be capable of withstanding a pressure of 2068.5 kPa when held in a closed position.

Questions

1. State the factors that decide the pressure exerted by a liquid.

2. Calculate the pressure in kPa acting on a valve when the head of water above the valve is 15 m.

Answer: 147.15 kPa

3. Calculate the force acting upon the base of a tank measuring 3 m by 2 m by 2 m deep. The tank contains water to a depth of 1.6 m.

Answer: 94.176 kN

4. Calculate the force acting upon a 76 mm diameter gate valve and the force acting upon the gate when the head of water above the valve is 20 m.

Answers: 196.2 kPa; 0.89 kN or 890 N

5. Calculate the force acting upon the top of a vertical calorifier when its diameter is 1.5 m and the head of water above the top is 5 m.

Answer: 86.678 kN or 86 678 N

6. Define the term 'centre of pressure'.

7. Describe how the total force acting upon one vertical side of a tank is calculated.

8. State why the metal struts placed across a tank full of water are placed near the top or one-third of the height from the base.

9. A pressure gauge on the top of a boiler gives a reading of 150 kPa. Calculate the head of water above the gauge.

Answer: 15.29 m

10. Define the terms 'absolute pressure' and 'gauge pressure'.

11. Calculate the 'absolute pressure' acting on the base of a tank when the head of water above the base is 12 m.

Answer: 117.72 kPa + 101 kPa = 218.72 kPa

12. A water-filled manometer is connected to a ventilating duct to measure the air pressure inside the duct. If the reading shows a head of water of 60 mm, calculate the air pressure in pascals.

Answer: 588.6 Pa

13. A water-filled manometer is connected to a gas pipe to measure the pressure of gas in the pipe. If the reading shows 150 mm head of water calculate the gas pressure in millibars.

Answer: 14.715 millibars

14. A rectangular tank measures 4 m long by 3 m wide by 2 m deep and contains water to a depth of 1.5 m.
    (a) Find, by means of a pressure diagram, the resultant force acting on one vertical end.
    (b) Show where the resultant force acts.

15. A pumping station is below ground level and the ground water rises to a maximum height, above the underside of the floor, of 1.5 m. If the floor has an area of 6 m², calculate the total force of water acting on the underside of the floor.

Answer: 88.29 kN

16. Water in a main has a pressure of 500 kPa and is to be used to supply a cold-water storage cistern having a ballvalve inlet 30 m above the centre of the main. Calculate the pressure acting on the ballvalve.

Answer: 205.7 kPa

17. A range of W.C. flushing cisterns are to be supplied from a cold-water storage cistern. The height from the water surface in the cistern and the ballvalves is 80 m. What type of ballvalve should be specified according to pressure classification?

Answer: high pressure (784.8 kPa)

18. A hydraulic press has pistons having areas of 500 mm² and 1500 mm² respectively. What load can be lifted on the larger piston if a force of 15 kN is applied to the smaller one?

Answer: 45 kN lifting force; 4.586 tonnes load lifted

Chapter 3

Hydraulics

Hydraulics

The subject of hydraulics concerns the study of fluids in motion. In order to maintain a certain rate of flow of fluid through a pipe or channel, a force is required to overcome the friction that exists between the surface of the pipe or channel and the moving fluid. In pipes this resistance is proportional to:

1. The square of the velocity (V^2).
2. The area of the surface in contact with the fluid.

   In the case of a pipe flowing full, this is equal to πdL, the curved surface of a cylinder,

   where d = diameter of pipe
   L = length of pipe
3. The nature of the internal surface.

   A rough surface will offer more resistance to flow than a smooth surface. Other factors that have an effect on flow are the density and viscosity of the fluid. If all the factors are constant, it has been shown by experiment that the discharge through a pipe is directly proportional to the square root of the fifth power of the diameter ($\sqrt{d^5}$).

   This is one of the discharge laws used in the Box formula, which will be described later in this chapter.

Relative discharging power of pipes

The relative discharging power of pipes are as the square root of the fifth power of their diameters.

$$N = \sqrt{\left(\frac{D}{d}\right)^5}$$

where N = number of branch pipes
D = diameter of main pipe
d = diameter of branch pipes

Example 3.1. *Calculate the diameter of a water main to supply twenty 32 mm diameter short branch pipes.*

$$N = \sqrt{\left(\frac{D}{d}\right)^5}$$

$$N^2 = \left(\frac{D}{d}\right)^5$$

$$\sqrt[5]{N^2} = \frac{D}{d}$$

$$D = \sqrt[5]{N^2} \times d$$

$$= \sqrt[5]{20^2} \times 32$$

$$= 106 \text{ mm}$$

The nearest commercial pipe size is 100 mm but this is undersized, so a 125 mm diameter pipe might be chosen.

Example 3.2. *Calculate the diameter of a water main to supply thirty 19 mm diameter short branch pipes.*

$$D = \sqrt[5]{N^2} \times d$$

$$= \sqrt[5]{30^2} \times 19$$

$$= 74 \text{ mm}$$

The nearest commercial pipe size is 76 mm, which would be satisfactory.

Example 3.3. *Calculate the number of 19 mm diameter short branch pipes that may be supplied from a 50 mm diameter main pipe.*

$$N = \sqrt{\left(\frac{D}{d}\right)^5}$$

$$= \sqrt{\left(\frac{50}{19}\right)^5}$$

$$= 11.24$$

Eleven branch pipes could be supplied.

Relative capacities of pipes

Although it is not a problem of hydraulics, it is sometimes necessary to compare the amount of water contained in pipes of various diameters. The relative capacities of pipes are as the square of their diameters.

$$N = \frac{D^2}{d^2}$$

where N = number of smaller pipes
$\qquad d$ = diameter of smaller pipes
$\qquad D$ = diameter of main pipe

Example 3.4. *How many 25 mm diameter branch pipes will be required to hold the same volume of water as one 76 mm diameter pipe?*

$$N = \frac{D^2}{d^2}$$

$$= \frac{76^2}{25^2}$$

$$= 9.243$$

Nine 25 mm diameter branch pipes would be required.

The relative discharging power of pipes takes into account the greater frictional loss in the branch pipes compared with the main pipe. A comparison of relative discharging power of pipes and relative capacities may therefore be made.

Example 3.2 showed that a 74 mm diameter main pipe would be required to supply thirty 19 mm diameter branch pipes. A comparison of relative capacities of pipes may be made using this example.

Example 3.5. *Calculate the number of 19 mm diameter pipes required to hold the same volume of water as one 74 mm diameter pipe.*

$$N = \frac{D^2}{d^2}$$

$$= \frac{74^2}{19^2}$$

$$= 15.17$$

It will be seen that about fifteen 19 mm branch pipes have the same capacity as one 74 mm diameter main pipe, whereas thirty 19 mm branch pipes have the same discharging power as one 74 mm diameter main pipe.

Types of flow (see Fig. 3.1)

When a fluid such as water flows in a pipe or channel, the flow may be either streamline or turbulent.

Streamline or laminar flow occurs when the motion of each molecule of fluid is in a straight line in the same direction as the axis of the pipe or channel.

Turbulent flow occurs when the motion of each molecule of the fluid is in a disorderly or disturbed manner.

Osborne Reynolds in 1883 recorded a number of experiments to determine the laws of resistance in pipes and whether the flow was streamline or turbulent. By introducing a filament of dye into the flow of water along a glass pipe, he showed that at low velocities the filament appeared in a straight line, indicating streamline flow. At high velocities, the filament, after passing a little way along the pipe, suddenly mixed with the surrounding water, indicating that the flow had become turbulent. The velocity of flow where the motion changes from streamline to turbulent is termed 'critical velocity'.

Fig.3.1 Streamline and turbulent flow

Reynolds number

Experiments with pipes of different diameters and with water at different temperatures led Reynolds to conclude that the type of flow was determined by:

(a) density of the fluid
(b) velocity of flow
(c) diameter of pipe
(d) viscosity of the fluid

Reynolds number $(R) \dfrac{\rho V d}{\nu}$

where ρ = density of fluid in kg/m^3
  V = velocity in m/s
  d = diameter of pipe in metres
  ν = viscosity of the fluid in Pa s

Whatever the type of fluid or its temperature, if R is less than 2000 the flow will be streamline. If R is greater than 2800 the flow will be turbulent.

Example 3.6. *Calculate the Reynolds number for a fluid flowing through a 25 mm bore pipe at 2 m/s when its density is 900 kg/m^3 and its viscosity is 0.015 Pa s.*

$$R = \frac{\rho V d}{\nu}$$

$$= \frac{900 \times 2 \times 0.025}{0.015}$$

$$= 3000$$

This number would give turbulent flow which is commonly found in practical examples.

Streamline flow, however, prevents waste of energy and is therefore preferable in this respect. Higher velocities are used in order to reduce the diameters of pipes.

Pipe flow formula

The basic formula for the flow of water through pipes usually known as the D'arcy formula is given in the following expression:

$$b = \frac{4 f L V^2}{2 g D}$$

where b = loss of head due to friction in metres
  f = coefficient of friction
  L = length of pipe in metres
  V = velocity of flow in metres per second
  g = acceleration due to gravity in metres per second squared
  D = diameter of pipe in metres

The value of the coefficient of friction depends upon the condition of the internal bore of the pipe. A value of 0.007 may be used for most problems and this is the value chosen by Thomas Box in his well-known formula which was derived from the D'arcy formula. The value may be found by experiment.

Example 3.7. *Find the loss of head due to friction in a 50 mm diameter pipe 30 m long when the velocity of flow is 2 m/s.*

$f = 0.007$
$g = 9.81$

$$b = \frac{4 f L V^2}{2 g D}$$

$$= \frac{4 \times 0.007 \times 30 \times 2^2}{2 \times 9.81 \times 0.05}$$

$$= 3.425 \text{ m}$$

Example 3.8. (see Fig. 3.2). *During an experiment conducted on a 25 mm bore copper pipe it was observed that the loss of head recorded by two water-filled manometers was 100 mm head of water. The tappings were 2 m apart and the velocity of flow 1 m/s. Calculate the coefficient of friction.*

Fig.3.2 Loss of head due to friction

$$b = \frac{4 f L V^2}{2 g D}$$

by transposition

$$f = \frac{b \times 2g \times D}{4 L V^2}$$

$$= \frac{0.1 \times 2 \times 9.81 \times 0.025}{4 \times 2 \times 1^2}$$

$$= 0.006\,131$$

Example 3.9. *Find the discharge in litres per second through a 32 mm bore pipe when the head lost due to friction is 5 m, length of pipe 20 m and the coefficient of friction is 0.007.*

$$b = \frac{4 f L V^2}{2 g D}$$

29

In order to find the discharge through the pipe, the velocity will be required.

By transposition

$$V = \sqrt{\frac{b \, 2 \, gD}{4 \, fL}}$$

$$= \frac{5 \times 2 \times 9.81 \times 0.032}{4 \times 0.007 \times 20}$$

$$= \sqrt{\frac{3.139}{0.56}}$$

$$= 2.367 \text{ m/s}$$

$Q = VA$

where Q = flow rate in m³/s
     V = velocity of flow in m/s
     A = area of pipe in m²

$Q = V \times 0.7854 \times D^2$
    $= 2.367 \times 0.7854 \times 0.032 \times 0.032$
    $= 0.0019 \text{ m}^3/\text{s}$

discharge = 1.9 litre/s

Thomas Box formula

This is a well-known formula for pipe sizing given in the following expression:

$$q = \sqrt{\frac{d^5 \times H}{25 \times L \times 10^5}}$$

where q = discharge through pipe in litres per second
     d = diameter of pipe in millimetres
     H = head of water in metres
     L = total length of pipe in metres

As mentioned earlier, the formula was derived from the D'arcy formula in which the coefficient of friction 'f' chosen by Box was 0.007. This value is reasonably reliable for smooth pipes such as copper, plastic, stainless steel and lead. The coefficient of friction for galvanised mild steel pipes, however, is higher and therefore the Box formula is unreliable for these pipes.

Note: Steel pipes are graded as class A, B and C. Class A is a light grade and is not used for water services, but class B is used extensively. The internal diameter of class B (medium grade) is slightly greater than the internal diameter of a similar copper or plastic pipe. This is to permit threading of the pipe using a standard British pipe thread. This increase in diameter of class B pipe allows for a better flow of water than would otherwise be obtained and compensates for the greater resistance to flow of water. Class C pipe is heavy grade and is often used for underground water services.

Example 3.10 (see Fig. 3.3). *Two cold-water storage cisterns each containing 4500 litres are sited in a cistern room on the roof of a building and require to be refilled every two hours. The vertical height from the water main and the* *ballvalves is 26 m and the horizontal lengths from this main and the ballvalves is 9 m. If the pressure on the main supplying the cisterns is 300 kPa during the peak demand periods, calculate the diameter of the rising main assuming that the frictional losses on the pipe and fittings to be 30 per cent on the net length of pipe.*

Fig.3.3   Size of rising main

$$\text{discharge in litres per second} = \frac{4500 \times 2}{2 \times 60 \times 60}$$

$$= 1.25$$

Head of water at ballvalves in metres (residual head)

pressure on water main = 300 kPa

$$\text{head of water on main} = \frac{300}{9.81}$$

$$= 30.58 \text{ m}$$

head lost due to height
to ballvalves   = 26 m
∴ residual head on ballvalves = 30.58 − 26
$$= 4.58 \text{ m}$$

Length of pipe allowing 30 per cent for frictional losses

height = 26
horizontal length = 9
                —————
                 35
add 30 per cent
frictional resistance = 10.5
                    —————
total length of pipe = 45.5 m

Values required in the Box formula:

1. discharge litre/s = 1.25
2. head of water in metres = 4.58
3. length of pipe in metres = 45.5

$$q = \sqrt{\frac{d^5 \times H}{25 \times L \times 10^5}}$$

$$q^2 = \frac{d^5 \times H}{25 \times L \times 10^5}$$

by transposition $d^5 = \dfrac{q^2 \times 25 \times L \times 10^5}{H}$

$$d = \sqrt[5]{\frac{q^2 \times 25 \times L \times 10^5}{H}}$$

by substitution $d = \sqrt[5]{\dfrac{1.25^2 \times 25 \times 45.5 \times 10^5}{4.58}}$

$$= 32.94 \text{ mm}$$

$$= 33 \text{ mm (approx.)}$$

The nearest commercial pipe size is 32 mm but this is slightly undersized; the next size is 38 mm which is 5 mm oversized; therefore a 32 mm diameter rising main would normally be chosen.

From an inspection of the Box formula it will be seen that the following pipe discharge laws will hold.

1. H varies directly with q^2.
2. H varies directly with L.
3. q varies directly with $\sqrt{H}$.
4. q varies directly with $\sqrt{d^5}$.

Variation of discharge

As the loss of head varies directly with the square of the discharge, then·the following law of proportionality will hold:

$$H_1 \times q_1^2 = H_2 \times q_1^2$$
where H_1 = original head
H_2 = new head
q_1 = original discharge
q_2 = new discharge

Example 3.11. *If a 13 mm diameter pipe can discharge 0.15 litres per second with a head of water of 4 m, calculate the head of water required when discharging 0.3 litres per second.*

$$H_1 \times q_2^2 = H_2 \times q_1^2$$

$$H_2 = H_1 \times \left(\frac{q_2}{q^1}\right)^2$$

$$= 4 \times \left(\frac{0.3}{0.15}\right)^2$$

$$= 16 \text{ m}$$

Example 3.12. *If a 38 mm diameter pipe discharges 2 litres per second, calculate the discharge if the pipe run is increased to 76 mm diameter.*

$$q_1 \times \sqrt{d_2^5} = q_2 \times \sqrt{d_1^5}$$

$$q_2 = q_1 \times \sqrt{\left(\frac{d_2}{d_1}\right)^5}$$

$$= 2 \times \sqrt{\left(\frac{76}{38}\right)^5}$$

$$= 2 \times \sqrt{2^5}$$

$$= 11.314 \text{ litre/s}$$

$$= 11.3 \text{ litre/s (approx.)}$$

Example 3.13. *If a 13 mm diameter branch pipe is required to discharge 0.15 litres per second, calculate the size of main required to serve ten such branch pipes.*

total discharge = $10 \times 0.15 = 1.5$ litre/s

$$q_1 \times \sqrt{d_2^5} = q_2 \times \sqrt{d_1^5}$$

$$d_2 = \sqrt[5]{\left(\frac{q_2}{q_1}\right)^2} \times d_1$$

$$= \sqrt[5]{\left(\frac{1.5}{0.15}\right)^2} \times 13$$

$$= \sqrt[5]{10^2} \times 13$$
$$= 2.512 \times 13$$
$$= 32.656 \text{ mm}$$
$$= 33 \text{ mm (approx.)}$$

Example 3.14.   *Calculate the discharge in litres per second through a 25 mm diameter pipe when the total length is 30 m and there is a constant head of 8 m.*

$$q = \sqrt{\frac{d^5 \times H}{25 \times L \times 10^5}}$$

$$= \sqrt{\frac{25^5 \times 8}{25 \times 30 \times 10^5}}$$

$$= 1.021 \text{ litre/s}$$
$$= 1 \text{ litre/s (approx.)}$$

Example 3.15.   *Calculate the head of water necessary to provide a discharge of 0.2 litre/s through a 13 mm diameter pipe 6 m long.*

$$q = \sqrt{\frac{d^5 \times H}{25 \times L \times 10^5}}$$

by transposition $H = \dfrac{q^2 \times 25 \times L \times 10^5}{d^5}$

$$= \frac{0.2^2 \times 25 \times 6 \times 10^5}{13^5}$$

$$= 1.616 \text{ m}$$
$$= 1.6 \text{ m (approx.)}$$

Example 3.16.   *Calculate the diameter of a pipe to discharge 1.25 litre/s when the constant head is 4 m and the total length of pipe 45.5 m.*

$$q = \sqrt{\frac{d^5 \times H}{25 \times L \times 10^5}}$$

by transposition $d = \sqrt[5]{\dfrac{q^2 \times 25 \times L \times 10^5}{H}}$

$$= \sqrt[5]{\frac{1.25^2 \times 45.5 \times 25 \times 10^5}{4}}$$

$$= 33.85 \text{ mm}$$
$$= 34 \text{ mm (approx.)}$$

Example 3.17.   *Calculate the discharge in litres per second through a 32 mm diameter pipe when the head is 5 m and the total length 20 m.*

$$q = \sqrt{\frac{d^5 \times H}{25 \times L \times 10^5}}$$

$$= \sqrt{\frac{32^5 \times 5}{25 \times 20 \times 10^5}}$$

$$= 1.8 \text{ litre/s (approx.)}$$

Residual head (see Fig. 3.4)

The head of water calculated by the Box formula will provide the necessary pressure of water to overcome the resistance of water flowing through a pipe at a certain rate.

In practice, a pressure is also required to be left at the discharging point in order to operate certain fittings such as showers, sprinklers and flushing valves. The head of water left after overcoming friction in the pipe is known as the residual head and manufacturers will provide the required details for their fittings.

If in example 3.15 the residual head required is 1 m, then the total head would be:

head required to overcome + head required to operate
        friction   the valve

=   1.616   +   1.000

= 2.616 m total head or effective head

Fig.3.4   Residual head on valve

Resistance of fittings

The resistance to the flow of water due to fittings such as bends, elbows, tees, valves, etc. may be expressed in head of water.

The fundamental equation for velocity head given in Chapter 2 is

$$b = \frac{V^2}{2g}$$

Expressing the resistance of fittings in terms of velocity head the following may be used:

$$b = k \left[\frac{V^2}{2g} \right]$$

where b = velocity head lost due to fittings in metres
       k = a number determined experimentally
       V = velocity in m/s (velocity-head factor)
       g = 9.81 m/s^2

Expressing a fitting in terms of velocity head,

$$k = \frac{b \, 2g}{V^2}$$

k is not constant for any given fitting but varies with the velocity of flow. Providing a mean velocity is used when determining k, the variation above or below this velocity does not detract from the usefulness of this method of expressing resistance.

Resistance expressed as equivalent length of pipe

This method of allowing for resistance in terms of equivalent length of pipe is widely used by designers. These lengths are added to the net length of pipe to give the effective length used in either the D'arcy or the Box formula.

Table 3.1 gives values of k and equivalent lengths in pipe diameters for some common fittings and connections.

Table 3.1

Type of fitting or connection	Approx. value of k	Approx. equivalent length in pipe diameters
90° Elbow	1.00	30–36
Easy bend	0.3	10
Flush connection to cisterns, tanks, etc.	0.5	20
Globe valve	10	340
Tee		
straight	0.5	20
reducing one side	0.75	30
reducing two sides	1.00	36
water entering branch	2–2.5	70–90
Gate valve	0.2	7

Example 3.18. *Calculate the head lost due to friction through a globe valve when the velocity of flow is 2 m/s.*

$$b = k \frac{V^2}{2g}$$

$$= 10 \times \frac{2^2}{2 \times 9.81}$$

$$= 2.039 \text{ m}$$

Example 3.19. *Find the effective length of pipe when a 25 mm diameter pipe has the following fittings in its run, 2 globe valves, 4 bends and 2 straight tees. Net length of pipe 50 m.*

Resistance due to fittings:

globe valves = 0.025 × 2 × 340 = 17 m

bends   = 0.025 × 4 × 10 =   1 m

tees   = 0.025 × 2 × 20 =   1 m

         total   19 m

effective length = 50 + 19 = 69 m

The Bernoulli theorem (see Fig. 3.5)

The theorem states that the total energy of each particle of a fluid is the same provided that no energy enters or leaves the system at any point. The theorem therefore is an aspect of the law of conservation of energy and if there is a loss of one type of energy there must be a corresponding gain in the other or vice versa.

Fig. 3.5 The Bernoulli theorem

$$\text{total energy } H = z + \frac{p}{\rho g} + \frac{V^2}{2g} = \text{constant}$$

where z = potential energy due to the position of the particles (m)

$\dfrac{p}{\rho g}$ = pressure energy due to the depth of the particles below the fluid (m)

$\dfrac{V^2}{2g}$ = kinetic energy due to the movement of the particles of the fluid (m)

In a closed system the total energy at only two points must be equal providing there are no gains or losses.

$$z_1 + \frac{p_1}{\rho g} + \frac{V_1^2}{2g} = z_2 + \frac{p_2}{\rho g} + \frac{V_2^2}{2g}$$

The Venturi meter (see Fig. 3.6)

A Venturi meter consists of a short inlet taper cone which reduces the pipe diameter to the throat diameter. This is followed by a divergent section which for minimum frictional loss should be longer than the inlet cone.

A 'U' tube is used to measure the pressure difference between the inlet and the throat.

Fig.3.6   The Venturi meter

Applying the Bernoulli equation:

$$z_1 + \frac{p_1}{\rho g} + \frac{V_1^2}{2g} = z_2 + \frac{p_2}{\rho g} + \frac{V_2^2}{2g}$$

For a horizontal meter $z_1 = z_2$ and are cancelled out of the two sides of the equation.

therefore $\dfrac{p_1}{\rho g} + \dfrac{V_1^2}{2g} = \dfrac{p_2}{\rho g} + \dfrac{V_2^2}{2g}$

and $\dfrac{V_2^2 - V_1^2}{2g} = \dfrac{p_1 - p_2}{\rho g}$

Since the same amount of fluid must pass through sections 1 and 2

$$a_1 V_1 = a_2 V_2$$

where a_1 = cross-sectional area of section 1

a_2 = cross-sectional area of section 2

By substitution and the addition of a coefficient of discharge, the expression for the flow in a Venturi meter can be shown:

$$Q = Cd\, a_1 \sqrt{\dfrac{2gH}{(a_1/a_2)^2 - 1}}$$

where Q = discharge in m³/s

Cd = coefficient of discharge

a_1 = area of entrance pipe in m²

a_2 = area of throat in m²

H = differential head of water between inlet and throat in metres

---

Example 3.20. *A 225 mm diameter water main contains a Venturi meter whose throat diameter is 76 mm. A 'U' tube manometer containing mercury shows a difference in head between the inlet and the throat of 200 mm. If the coefficient of discharge is 0.95, calculate the rate of flow through the meter in m³/s.*

Since mercury is approximately 13.6 times heavier than water, the head of water equivalent to 200 mm will be:

head of water in mm = 13.6 × 200

= 2720

head of water in metres = 2.72

The areas of the inlet and throat may be found from $0.7854\, D^2$

$$a_1 = 0.7854 \times 0.225^2$$
$$= 0.0397 \text{ m}^2$$
$$a_2 = 0.7854 \times 0.076^2$$
$$= 0.0045 \text{ m}^2$$
$$H = 2.72$$
$$g = 9.81$$
$$Cd = 0.95$$

discharge $= Cd\, a_1 \sqrt{\dfrac{2gH}{(a_1/a_2)^2 - 1}}$

$$= 0.95 \times 0.0397 \sqrt{\dfrac{2 \times 9.81 \times 2.72}{(0.0397/0.0045)^2 - 1}}$$

$$= 0.0377 \sqrt{\dfrac{53.366}{76.83}}$$

$$= 0.0377 \times 0.833$$

$$= 0.0314 \text{ m}^3/\text{s or } 31.4 \text{ litre/s}$$

39

Questions

1. Calculate by the relative discharging power formula the number of 13 mm diameter short branch pipes that may be supplied from a 25 mm diameter main pipe.

Answer: five 13 mm diameter pipes

2. Calculate the diameter of a main pipe to supply thirty 19 mm diameter short branch pipes.

Answer: 74 mm diameter

3. Describe streamline and turbulent flow. What is 'critical velocity' of flow?

4. Describe an experiment to find the coefficient of friction for the flow of water through a pipe.

5. Using the D'arcy formula, calculate the diameter of a pipe in millimetres when the length of pipe is 15 m, velocity of flow 3 m/s and the head lost due to friction 4 m. Use a coefficient of friction of 0.005 and $g = 9.81$.

Answer: 34.42 mm

6. If a 19 mm diameter pipe can discharge 4 litres per second with a head of water of 8 m, calculate the head of water when discharging 2 litres per second.

Answer: 2 m

7. If a 50 mm diameter pipe can discharge 8 litres per second, calculate the discharge if the pipe is reduced to 25 mm diameter.

Answer: 1.4 litre/s

8. Using the Box formula, calculate the diameter of a pipe in millimetres when the discharge required is 1.5 litres per second. The total length of pipe is 20 m and there is a constant head of water of 20 m.

Answer: 20.65 mm (a 25 mm diameter pipe would be used)

9. Using the Box formula, calculate the head of water required when the discharge is 5 litres per second through a 50 mm diameter pipe having a total length of 45 m.

Answer: 9 m

10. Calculate the diameter of a rising main using the following factors:
    (a) discharge required at the ballvalve 1.5 litres per second
    (b) pressure on the main during the peak demand period 300 kPa
    (c) height from the water main to the ballvalve 10 m
    (d) total length of pipe 20 m

Answer: 21 mm (a 25 mm diameter pipe would be suitable)

11. Define the term 'equivalent pipe length'.

12. Calculate the loss of head due to friction through fittings when a pipe run

contains 2 easy bends, 2 straight tees, and 1 gate valve. The velocity of flow is 2 metres per second.

Answer: 0.122 + 0.24 + 0.048 = 0.41 m

13. Define Reynolds number and state the factors required for its evaluation.

14. Define velocity head and residual head.

15. Explain the Bernoulli theorem and describe how the theorem may be used to measure the flow of water through a pipe.

16. A 300 mm diameter water main contains a Venturi meter whose throat diameter is 100 mm. A 'U' tube manometer containing mercury shows a difference in head between the inlet and the throat of 150 mm. If the coefficient of discharge is 0.98, calculate the rate of flow through the meter in m^3/s.

Answer: 0.0486 m^3/s or 48.6 litre/s

Chapter 4

Flow in drains and open channels, sewage disposal

Drains and open channels

Chezy formula

Various formulae, tables and charts may be used to find the gradient or velocity of flow of water in a drain or channel. One of the best-known formulae which may be used for pipes or channels is that due to Chezy, expressed as follows:

$$V = C \sqrt{m\, i}$$

where C = Chezy constant
V = velocity of flow (m/s)
m = hydraulic mean depth (m)
i = inclination or fall

The Chezy constant may be found from the following formula:

$$C = \sqrt{\frac{2g}{f}}$$

where g = acceleration due to gravity
f = coefficient of friction

For drainage, the average coefficient of friction may be taken as 0.0064 and therefore C would be given by

$$C = \sqrt{\frac{2 \times 9.81}{0.0064}}$$

= 55 (approx.)

$$m = \frac{\text{cross-sectional area of flow}}{\text{length of wetted perimeter}}$$

For drains flowing half- or full-bore, the hydraulic mean depth is equal to $D \div 4$, which can be shown as follows:

full bore $\qquad m = \frac{\pi}{4} D^2 \div \pi D$

$$= \frac{\overset{1}{\cancel{\pi}}}{4} \times \frac{\overset{D}{\cancel{D^2}}}{1} \times \frac{1}{\underset{1}{\cancel{\pi}}} \times \frac{1}{\underset{1}{\cancel{D}}}$$

by cancellation $m = \dfrac{D}{4}$

half-bore $\qquad m = \frac{\pi}{8} D^2 \div \frac{\pi D}{2}$

$$= \frac{\overset{1}{\cancel{\pi}}}{\underset{4}{\cancel{8}}} \times \frac{\overset{D}{\cancel{D^2}}}{1} \times \frac{\overset{1}{\cancel{2}}}{1} \times \frac{1}{\underset{1}{\cancel{\pi}}} \times \frac{1}{\underset{1}{\cancel{D}}}$$

by cancellation $m = \dfrac{D}{4}$

Figure 4.1 shows the relationship between the cross-sectional area of flow and the length of the wetted perimeter for half- and full-bore discharge.

Fig.4.1 Hydraulic mean depth for full-bore or half-bore

The inclination, fall or gradient is equal to $H \div L$, where (L) is the length of drain in metres with a head (H) of 1 m (see Fig. 4.2).

Fig.4.2 Inclination, fall or gradient

The Chezy formula was derived from the D'arcy formula described in Chapter 3. The coefficient of friction of 0.0064 is comparable to that used by Box in his pipe-sizing formula.

Example 4.1. *Using the Chezy formula, calculate the discharging capacity of a 150 mm diameter drain flowing full when laid to a fall of 1 in 140. (Take C = 55.)*

$$V = C \sqrt{m}\, i$$

$$= C \sqrt{\frac{D}{4} \times \frac{1}{L}}$$

$$= 55 \sqrt{\frac{0.15}{4} \times \frac{1}{140}}$$

$$= 55 \sqrt{0.016\,36}$$

$$= 0.9 \text{ m/s}$$

$$Q = V A$$

where Q = discharge in m³/s
$\quad V$ = velocity of flow in m/s
$\quad A$ = area of pipe

$$\therefore \quad Q = V \times \frac{\pi}{4} D^2$$

$$= 0.9 \times \frac{3.142}{4} \times 0.15 \times 0.15$$

$$= 0.0159 \text{ m}^3\text{/s or } 15.9 \text{ litre/s}$$

Crimp and Bruges formula

This is another well-known formula often used by engineers, it is expressed as:

$$V = 84\, m^{2/3}\, i^{1/2}$$

or $\qquad V = 84 \sqrt[3]{m^2}\ \sqrt{i}$

where V = velocity of flow in m/s
$\quad m$ = hydraulic mean depth
$\quad\ i$ = inclination or fall

The formula may be used to calculate the velocity of flow in the 150 mm diameter drain laid at a gradient of 1 in 140 given in example 4.1.

$$V = 84 \sqrt[3]{m^2}\ \sqrt{i}$$

$$= 84 \sqrt[3]{\left[\frac{D}{4}\right]^2}\ \sqrt{\frac{1}{L}}$$

$$= 84 \sqrt[3]{\left[\frac{0.15}{4}\right]^2}\ \sqrt{\frac{1}{140}}$$

$$= 84 \sqrt[3]{0.0014} \times 0.0845$$

$$= 84 \times 0.1119 \times 0.0845$$

$$= 0.794 \text{ m/s}$$

$$= 0.8 \text{ m/s (approx.)}$$

Example 4.2. *Calculate the gradient required for a 100 mm diameter drain to run half-full at an average velocity of flow of 1.2 m/s. (Take C = 55.)*

$$V = C \sqrt{m \, i}$$

and $\quad V = C \sqrt{\dfrac{D}{4} \times \dfrac{1}{L}}$

by transposition

$$\left[\frac{V}{C} \right]^2 = \frac{D}{4} \times \frac{1}{L}$$

$$\frac{1}{L} = \left[\frac{V}{C} \right]^2 \times \frac{4}{D}$$

$$L = \left[\frac{C}{V} \right]^2 \times \frac{D}{4}$$

by substituting the values given in the question

$$L = \left[\frac{55}{1.2} \right]^2 \times \frac{0.1}{4}$$

$$= 2100.69 \times 0.025$$

$$= 52.52 \text{ (approx.)}$$

A gradient of 1 in 52.5 would be required.

Note: For flows in channels having other than circular cross-sections, the hydraulic mean depth may also be calculated.

Example 4.3. *Figure 4.3 shows a cross-section of a rectangular channel. Calculate the gradient if the average velocity of flow is to be 0.8 m/s.*

Area of flow = bd

Wetted perimeter = b + (2d)

d = 300 mm

b = 450 mm

Fig.4.3 Hydraulic mean depth for a rectangular channel

---

the hydraulic mean depth $= \dfrac{b \, d}{b + (2d)}$

$$= \frac{0.45 \times 0.3}{0.45 + (2 \times 0.3)}$$

$$= \frac{0.135}{0.45 + 0.6}$$

$$= \frac{0.135}{1.05}$$

$$= 0.1286$$

$$V = C\sqrt{m\,i}$$

$$= C\sqrt{m \times \frac{1}{L}}$$

by transposition $\left[\dfrac{V}{C}\right]^2 = m \times \dfrac{1}{L}$

$$\left[\frac{V}{C}\right]^2 \times \frac{1}{m} = \frac{1}{L}$$

$$L = \left[\frac{C}{V}\right]^2 \times m$$

$$= \left[\frac{55}{0.8}\right]^2 \times 0.1286$$

$$= 4726.5625 \times 0.1286$$

$$= 607.83$$

A gradient of 1 : 608 (approx.) is required.

Value of (m) for surface and foul-water drains

It is usual practice to design surface-water drains to run full-bore and foul-water drains to run at a maximum depth equal to three-quarters of the internal diameter of the pipe.

It has been shown that the hydraulic mean depth for full-bore discharge is equal to the diameter divided by four (or multiplied by 0.25).

For foul-water drains it will be necessary to find the hydraulic mean depth before the velocity of flow or the gradient may be found by the Chezy formula.

Example 4.4 (see Fig. 4.4). *Calculate the velocity of flow in a 100 mm diameter foul-water drain flowing at a depth equal to three-quarters of the diameter when the gradient is 1 in 60; Chezy constant 55.*

area of water flow = area of segment of circle plus area of triangle

area of segment of circle $= \dfrac{\theta}{360} \times \dfrac{\pi}{4} D^2$

$$= \frac{240}{360} \times \frac{3.142}{4} \times \frac{0.1}{1} \times \frac{0.1}{1}$$

$$= 0.005\ 24\ \text{m}^2 \text{ (approx.)}$$

area of triangle $= \dfrac{b\,b}{2}$

$$= \frac{0.086 \times 0.025}{2}$$

Area of water flow = area of segment + area of triangle

Fig.4.4 Hydraulic mean depth for $\frac{3}{4}$ depth

$$= 0.0011 \text{ m}^2 \text{ (approx.)}$$
$$\text{area of water flow} = 0.005\ 24 + 0.0011$$
$$= 0.006\ 34 \text{ m}^2$$

Wetted perimeter

$$\text{arc of circle} = \frac{\theta}{360} \times \pi D$$

$$= \frac{240}{360} \times 3.142 \times 0.1$$

$$= 0.2095 \text{ m (approx.)}$$

$$\text{hydraulic mean depth} = \frac{\text{area of water flow}}{\text{wetted perimeter}}$$

$$= \frac{0.006\ 34}{0.2095}$$

$$= 0.0303 \text{ (approx.)}$$

Using Chezy formula to find velocity,

$$V = C\sqrt{m\,i}$$

$$= 55\sqrt{0.0303 \times \frac{1}{60}}$$

$$= 55 \times 0.0224$$

$$= 1.236 \text{ m/s (approx.)}$$

Table 4.1 may be used to find the hydraulic mean depth for pipes with water flowing at various depths.

47

Table 4.1

Depth of flow	Hydraulic mean depth
Full or $\frac{1}{2}$ full bore	diameter $\times$ 0.25
$\frac{3}{4}$ depth of flow	diameter $\times$ 0.30
$\frac{2}{3}$ depth of flow	diameter $\times$ 0.29
$\frac{1}{3}$ depth of flow	diameter $\times$ 0.19
$\frac{1}{4}$ depth of flow	diameter $\times$ 0.15

Note: The approximate maximum velocity of flow of water through a drain is obtained when the pipe is flowing at a depth equal to four-fifths of the internal diameter. The minimum self-cleansing velocity of flow is 0.8 m/s, but there is no upper limit for self-cleansing velocity.

Design of surface water drainage

Rainfall intensity. For storms expected to occur once a year the following formulae may be used.

(a) $R = \dfrac{750}{(T + 10)}$

where T is 5 to 10 minutes

(b) $R = \dfrac{1000}{(T + 20)}$

where T is 20 to 100 minutes

In these formulae, R = rainfall per hour in mm
T = duration of storm in minutes

Note: The maximum run-off at the point of outgo does not occur until the end of the time of concentration, when the whole area is contributing to the discharge. Where a time of concentration is used, such a time must be added to T in the formulae.

Rainfall run-off

The amount of water which can be expected from any given surface depends upon the following factors:

1. the area of the surface;
2. the type of surface;
3. whether the surface is level or sloping;
4. the intensity of rainfall;
5. the seasonal conditions, i.e. rate of evaporation.

The following formula may be used to find the volume of water run-off:

$$Q = \frac{A\ P\ R}{60 \times 1000 \times 60}$$

where Q = run-off in m³/s
A = area drained in m²

P = impermeability factor
R = rainfall intensity in mm/hr

Table 4.2 gives the impermeability factors for various types of surfaces.

Table 4.2

Type of surfaces	Impermeability factor (average)
Watertight surfaces	0.90
Asphalt pavements in good order	0.875
Closely jointed wood or stone pavements	0.825
Macadam roadways	0.425
Lawns and gardens	0.15
Wooded areas	0.105

Example 4.5. *Calculate the diameter and gradient of a pipe required to carry the run-off from an area of one hectare of lawns and gardens using the following factors:*

1. duration of storm 5 minutes;
2. impermeability factor 0.15;
3. time of concentration 18 minutes;
4. full-bore discharge;
5. velocity of flow 0.8 m/s;
6. Chezy constant 55.

Note: 1 hectare = 10 000 m^2. The time of concentration is not usually used unless it exceeds about 15 minutes.

$$R = \frac{750}{T + t + 10}$$

where R = rainfall intensity (mm/hr)
$\quad\quad T$ = duration of storm in minutes
$\quad\quad t$ = time of concentration in minutes

$$\therefore R = \frac{750}{5 + 18 + 10}$$

$$= \frac{750}{33}$$

$$= 22.73 \text{ mm/hr}$$

$$Q = \frac{A\,P\,R}{60 \times 1000 \times 60}$$

$$= \frac{10\,000 \times 0.15 \times 22.73}{60 \times 1000 \times 60}$$

$$= \frac{34\,095}{3\,600\,000}$$

$$= 0.009\,47 \text{ m}^3/\text{s}$$

also $Q = VA$

$$= V \times \frac{\pi}{4} D^2$$

$$D = \sqrt{\frac{4Q}{V\pi}}$$

$$= \sqrt{\frac{4 \times 0.009\,47}{0.8 \times 3.142}}$$

$$= \sqrt{\frac{0.037\,88}{2.5136}}$$

$$= 0.123$$

A 125 mm diameter pipe would be satisfactory.

To find the gradient:

$$L = \left[\frac{C}{V}\right]^2 \times \frac{D}{4}$$

$$= \left[\frac{55}{0.8}\right]^2 \times \frac{0.125}{4}$$

$$= 4726.56 \times 0.031\,25$$

$$= 147.705$$

A gradient of 1 in 148 would be satisfactory.

If flooding is liable to cause damage or inconvenience, a rainfall intensity of 50 mm/hr is often used.

Example 4.6 (see Fig. 4.5). *Calculate the diameter and gradient of a pipe required to carry the run-off from an asphalt-covered car park measuring 50 m × 30 m using the following factors:*

1. impermeability factor 0.875;
2. full-bore discharge;
3. rainfall intensity 50 mm/hr;
4. velocity of flow 1.5 m/s;
5. Chezy constant 55.

$$Q = \frac{APR}{60 \times 1000 \times 60}$$

$$= \frac{50 \times 30 \times 0.875 \times 50}{60 \times 1000 \times 60}$$

$$= \frac{65\,625}{3\,600\,000}$$

$$= 0.018\,229 \text{ m}^3/\text{s}$$

$$= 0.0182 \text{ m}^3/\text{s (approx.)}$$

Alternatively, Q may be found from the following formula:

Fig.4.5 Plan of surface-water drainage for an asphalt covered car park

$$Q = \frac{\overset{\text{area to be}}{\text{drained (m}^2)} \times \overset{\text{rainfall}}{\text{intensity (m/h)}} \times \overset{\text{impermeability}}{\text{factor}}}{3600}$$

$$= \frac{50 \times 30 \times 0.05 \times 0.875}{3600}$$

$$= 0.0182 \text{ m}^3/\text{s (approx.)}$$

$$Q = VA; \quad Q = V\frac{\pi}{4}D^2; \quad D = \sqrt{\frac{4\,Q}{V\,\pi}}$$

$$D = \sqrt{\frac{4 \times 0.0182}{1.5 \times 3.142}}$$

$$= 0.125 \text{ m}$$

$$= 125 \text{ mm}$$

To find the gradient:

$$L = \left[\frac{C}{V}\right]^2 \times \frac{D}{4}$$

$$= \left[\frac{55}{1.5}\right]^2 \times \frac{0.125}{4}$$

$$= 1344.444 \times 0.031\,25$$

$$= 42$$

gradient = 1 in 42

Size of soakaway

A soakaway may be sized on the basis of a rainfall intensity of 15 mm per hour. Assuming that three soakaways are to be used to take the discharge from the car park in example 4.6, the size of each may be calculated as follows:

$$\text{volume of each soakaway} = \frac{50 \times 30 \times 0.015}{3}$$

$$= 7.5 \text{ m}^3$$

assuming each soakaway is square on plan and 2 m deep,

$$\text{volume} = \text{area} \times \text{depth}$$

$$\text{area} = \frac{\text{volume}}{\text{depth}}$$

$$\text{length of side} = \sqrt{\frac{\text{volume}}{\text{depth}}}$$

$$= \sqrt{\frac{7.5}{2}}$$

$$= 1.94 \text{ m}$$

$$\therefore \text{ size of each soakaway} = 2 \text{ m} \times 2 \text{ m} \times 2 \text{ m}$$

Alternatively, two soakaways may be used

$$\text{volume of each soakaway} = \frac{50 \times 30 \times 0.015}{2}$$

$$= 11.25 \text{ m}^2$$

Assuming each soakaway is square on plan and 2 m deep,

$$\text{length of one side} = \sqrt{\frac{11.25}{2}}$$

$$= 2.372 \text{ m}$$

$$\therefore \text{ size of each soakaway} = 2.372 \text{ m} \times 2.372 \text{ m} \times 2 \text{ m}$$

Foul-water drains

The diameter of a foul-water drain will depend upon the discharge from the sanitary fittings during the peak demand period. The following factors will have to be considered:

1. the number and types of sanitary fittings;
2. the possible peak frequency of use of the fittings;
3. the average duration of discharge of the fittings;
4. the volume of water discharged from the fittings.

Discharge-unit method

When the number and types of sanitary fittings are known, the diameter and gradient of a foul-water drain may be found by the discharge-unit method. By this method, each appliance is given a discharge-unit value which represents the discharge capacity and frequency of use of the fitting. The sum of all the discharge units of the sanitary fittings are found.

Table 4.3 gives a list of discharge units and Table 4.4 the number of discharge units to be allowed on 100 mm, 125 mm and 150 mm diameter horizontal drains and their gradients.

Example 4.7.   By use of the discharge-unit method, find the diameter and gradient of a horizontal drain to serve a five-storey office containing 8 W.Cs,

Table 4.3 Discharge-unit values BS 5572 1978

Type of sanitary fitting	Interval between uses			
	Domestic (min)	Public (min)	Peak flow (min)	Discharge unit value
Automatic washing machine	250			4
WC (9-litre cistern)	20			7
		10		14
			5	28
Sink	20			6
		10		14
			5	27
Wash basin	20			1
		10		3
			5	6
Bath	75			7
		30		18
			—	—
Shower	Add 0.1 litre per second per spray			
Urinal (stall or bowl)		20		0.3

Table 4.4 Maximum number of discharge units to be allowed on horizontal branches BS 5572 1978

Internal diameter of pipe (mm)	Fall		
	1 in 111	1 in 45.5	1 in 22
100	230	430	1050
125	780	1500	3000
150	2000	3500	7500

8 basins, 2 sinks, and 4 urinals on each floor. The W.Cs are to be provided with 9 litre flushing cisterns.

W.Cs	8 × 5 × 14 DUs =	560
Basins	8 × 5 × 3 DUs =	120
Sinks	2 × 5 × 14 DUs =	140
Urinals	4 × 5 × 0.3 DUs =	6
	Total	**826**

With reference to Table 4.4, a 150 mm diameter drain having a gradient of 1 in 111 would be suitable.

Simultaneous demand factor

The discharge-unit value takes into account the number of fittings discharging simultaneously compared with the total number of fittings installed (usually expressed as a percentage). The number of fittings discharging simultaneously, during the peak demand period, may be found by the application of the theory of probability and a simplified formula may be used, given by the expression:

$$m = np + 1.8 \ [2 \ np \ (1 - p)]^{0.5}$$

where m = number of fittings discharging simultaneously

$p = \dfrac{t}{T}$ known as the usage ratio

n = total number of fittings installed
t = time in seconds fittings are discharging
T = time in seconds between usage

Example 4.8. *Calculate the number of fittings discharging simultaneously if 100 fittings, each of which take 10 seconds to discharge their contents, are used at 400-second intervals. Calculate the simultaneous demand factor.*

$$p = \frac{t}{T}$$

$$\therefore p = \frac{10}{400} = 0.025$$

$$m = (100 \times 0.025) + 1.8 \ [2 \times 100 \times 0.025 \ (1 - 0.025)]^{0.5}$$

$$= 4.3 \times \sqrt{4.875}$$

$$= 4.3 \times 2.2$$

$$= 6.5$$

$$\text{simultaneous demand} \atop \text{factor (per cent)} \quad = \frac{\text{number of fittings discharging simultaneously}}{\text{number of fittings installed}} \times \frac{100}{1}$$

$$= \frac{6.5}{100} \times \frac{100}{1}$$

$$= 6.5 \text{ per cent}$$

If the simultaneous demand factor and the rate of discharge from the fittings are known, the diameter and the gradient of a foul-water drain may be calculated.

Table 4.5 gives the approximate rate of discharge from various sanitary fittings.

Example 4.9. *If the fittings given in example 4.8 consist of 40 W.Cs, 50 basins and 10 sinks with 38 mm diameter wastes, calculate the diameter and gradient of a drain to carry their discharges when the velocity of flow is to be 1 m/s and the drain is to flow at a maximum of half-full bore. Chezy constant = 55.*

Discharge from fittings (see Table 4.5):

Table 4.5 Rate of discharge from sanitary fittings

Type of fitting	Discharge (litres per second)
Basin	0.60
Bath	1.00
Sink (32 mm diameter waste)	0.50
Sink (38 mm diameter waste)	1.00
Urinal	0.15
WC (9 or 14 litre cistern)	2.30
Shower	1.00

W.Cs $40 \times 2.3 = 92$

Basins $50 \times 0.6 = 30$

Sinks $10 \times 1.0 = \underline{10}$

Total $\underline{132}$ litre/s

allowing 6.5 per cent

simultaneous demand factor $= \dfrac{132}{1} \times \dfrac{6.5}{100}$

$$\text{discharge} = 8.58 \text{ litre/s}$$
$$= 0.008\,58 \text{ m}^3/\text{s}$$

$$Q = V A$$

$$= V \times \frac{\pi}{8} D^2 \text{ (half-full bore)}$$

$$D = \sqrt{\frac{Q \times 8}{V \times \pi}}$$

$$= \sqrt{\frac{0.008\,58 \times 8}{1 \times 3.142}}$$

$$= 0.1478 \text{ m}$$
$$= 148 \text{ mm}$$

A 150 mm diameter pipe would be satisfactory.

To find the gradient:

$$L = \left[\frac{C}{V}\right]^2 \times \frac{D}{4}$$

$$= \left[\frac{55}{1}\right]^2 \times \frac{0.15}{4}$$

$$= 3025 \times 0.0375$$
$$= 113.44$$

A gradient of 1 in 113 would be satisfactory.

If the discharge-unit method is used to find the diameter and gradient for the drain given in example 4.9, these would be as follows:

W.Cs   40 × 14 DUs = 560
Basins   50 ×   3 DUs = 150
Sinks   10 × 14 DUs = 140
           Total   850

With reference to Table 4.4, a 150 mm diameter drain having a gradient of 1 in 96 would be suitable.

Foul-water sewers

Experience has shown that, for the design of foul-water sewers on the separate system of drainage, a basis of six times the average 24-hour flow, determined by the water consumption per head of population, generally provides an adequate factor of safety.

The approximate water consumptions are: 140; 180; 240 litres per person per day.

Example 4.10. *Calculate the diameter of a foul-water sewer suitable for 150 housing units. The average number of occupants per house is assumed to be four and the water consumption 240 litres per person per day. The sewer is to be designed to run half-full bore, with a velocity of flow of 1 m/s.*

$$\text{rate of flow per day} = 240 \times 150 \times 4$$
$$= 144\,000 \text{ litres}$$

average flow rate during a
six-hour period assumed half
the daily flow $= 72\,000$ litres

$$\text{average flow rate per hour} = \frac{72\,000}{6}$$
$$= 12\,000 \text{ litres}$$

$$\text{average flow rate per second} = \frac{12\,000}{60 \times 60}$$
$$= 3.333$$

maximum flow rate per second
assuming four times the average $= 3.333 \times 4$
$$= 13.333 \text{ litre/s}$$
$$= 0.0133 \text{ m}^3/s$$

$$Q = V A$$
$$= \frac{V \pi D^2}{8}$$
$$D = \sqrt{\frac{8 Q}{V \pi}}$$
$$= \sqrt{\frac{8 \times 0.0133}{1 \times 3.142}}$$
$$= 0.184 \text{ m}$$
$$= 184 \text{ mm}$$

A 200 mm diameter pipe would be satisfactory.

The approximate peak flow rates for three different water consumptions are 140, 180 and 240 litres per person per day. If 180 litres per person per day was chosen in example 4.10, the diameter of the sewer would be calculated in example 4.11.

Example 4.11.

$$\text{rate of flow per day} = 180 \times 150 \times 4$$
$$= 108\,000 \text{ litres}$$

average flow rate during
six-hour period assuming half
the daily flow $\qquad = 54\,000$ litres

$$\text{average flow rate per hour} = \frac{54\,000}{6}$$
$$= 9000 \text{ litres}$$

$$\text{average flow rate per second} = \frac{9000}{60 \times 60}$$
$$= 2.5$$

maximum flow rate per second
assuming four times the average $= 2.5 \times 4$
$$= 10 \text{ litre/s}$$
$$= 0.01 \text{ m}^3/\text{s}$$

$$D = \frac{8\,Q}{V\,\pi}$$

$$= \sqrt{\frac{8 \times 0.01}{1.2 \times 3.142}}$$

$$= 0.146 \text{ m}$$
$$= 146 \text{ mm}$$

A 150 diameter pipe would be satisfactory.

The CP 301, Building Drainage, considers that not more than, say, 20 housing units should be connected to a 100 mm nominal bore drain and not more than 150 housing units should be connected to a 150 mm diameter nominal bore drain. Example 4.11 shows that a 150 mm nominal bore pipe is suitable when the water consumption is taken as 180 litres per person per day and a velocity of flow is 1.2 m/s.

Drainage above ground

Terminal velocity in stacks

It was shown in Chapter 1 that if a body is falling freely near to the surface of the earth it has an acceleration due to gravity of 9.81 m/s². When water flows down a vertical drainage stack, however, the forces of gravity and friction acting on the flow, soon balance and a terminal velocity is reached (see Fig. 4.6).

The height of stack necessary for terminal velocity to be achieved depends upon the diameter of the pipe, amount of flow and smoothness of the internal

Fig. 4.6 Frictional resistances balancing gravitation of water

bore. The height is only likely to be equal to one storey. There is, however, no precaution necessary to reduce the velocity of flow in tall stacks.

Capacities of stacks

Owing to the terminal velocity attained in stacks there is a limit to the flow capacity of any given stack. In order to reduce water and air disturbance to a minimum, the upper limit of a stack loading should be about a quarter full. This amount of loading provides room for an air core to flow down the stack during discharges (see Fig. 4.7).

Figure 4.8 shows how the air core is maintained at branch connections to the stack.

Fig. 4.7 Relationship between flow of water and air core

A formula for the sizing of a vertical stack when flowing quarter full is expressed as follows:

$$q = K \, d^{\frac{8}{3}} \text{ or } K \sqrt[3]{d^8}$$

where K = a constant, 0.000 032
q = discharge capacity in litres per second
d = diameter of stack in mm

Example 4.12. *Calculate the diameter of a vertical stack to discharge 2 litres per second when flowing quarter full.*

Plan at W.C. branch   Plan at waste-pipe branch

Fig.4.8   Air core at branch connection

$$q = 0.000\ 032 \sqrt[3]{d^8}$$

by transposition $d = \sqrt[8]{\left[\dfrac{q}{0.000\ 032}\right]^3}$

$$= \sqrt[8]{\left[\dfrac{2}{0.000\ 032}\right]^3}$$

$$= 63\ mm$$

A 64 mm diameter stack would be suitable.

Discharge unit value method

This method has been described previously for horizontal drains but the method may also be used for vertical discharge stacks using the Table 4.3 to find the sum of the discharge units and then referring to Table 4.6 to find the diameter of the stack.

Table 4.6   Maximum number of discharge units to be allowed on vertical stacks CP 304 1968

Nominal internal diameter of pipe (mm)	Discharge units
50	10
65	60
75	200 (not more than 1 WC)
90	350
100	750
125	2500
150	5500

Example 4.13.   *By use of the discharge unit value method find the diameter of a vertical stack to take the discharges from the following sanitary fittings in a 10 storey office block, 120 W.Cs with 9 litre flush, 120 basins, 20 sinks, and 40 urinals.*

59

```
W.Cs      120 x 14 DUs  =  1680
Basins    120 x  3 DUs  =   360
Sinks      20 x 14 DUs  =   280
Urinals    40 x 0.3 DUs =    12
                  Total   2332
```

By reference to Table 4.6, a 125 mm diameter stack would be suitable.

Diameter of ventilating pipes

The diameters of branch ventilating pipes are given in Table 4.7. Longer lengths of ventilating pipes will, however, have to be larger.

Table 4.7 Diameter of branch ventilating pipes

Diameter of discharge stack (mm)	Diameter of branch ventilating pipe (mm)
Smaller than 75 mm	$\frac{2}{3}D$
75 mm to 150 mm	½D

Sewage disposal

Septic tanks

This is where the primary treatment takes place and sewage should be retained in the tank between 16 and 48 hours. The tank should be sized accordingly and the Code of Practice BS6297, Small Sewage Treatment Work 1983, gives the following formula for the total capacity of a septic tank where desludging is carried out at not more than 12-monthly intervals:

$$C = 180 N + 2000$$

where C = capacity of tank in litres
        N = number of people served (with a minimum of four)

The Code of Practice recommends that septic tanks which are to serve a population of up to 100 persons should be divided into two compartments. The first compartment should have a length of twice its breadth and should provide two-thirds of the total capacity. The second compartment should be square on plan and sized to hold the remaining one-third of the total capacity. The total length of the tank therefore, ignoring the dividing wall, will be three times the breadth. The depth of the tank should be not less than 1.5 m.

Biological filter

This is where the final treatment of the sewage takes place and where it is purified sufficiently to discharge the effluent into a river, stream or soakaway. It is filled with filter media and its volume should be 1 m³ per head for up to 10 persons, 0.8 m³ per head for up to 50 persons and 0.6 m³ per head for up to 300 persons. The volume should be increased by about 30 per cent when sink grinders are installed. Its depth should be 1.8 m.

Fig.4.9 Sewage treatment plant for 100 persons

Humus tank

This may be required for larger sewage treatment plants to remove the humus from the effluent of the biological filter. The humus is an inoffensive, earth-like material which, if not removed, may form banks at the river or stream.

The following formula for the capacity of a humus tank for retention times, varying from 24 to 8 hours for a population range 10 to 8, and from 8 to 5 hours for a population range 50 to 300 may be used.

$$C = 30N + 1500$$

where C = capacity of tank in litres
N = number of people served

Example 4.14. *Calculate the sizes of the septic tank, biological filter and humus tank for a sewage treatment plant to serve 100 persons.*

Dimensions of septic tank

total capacity = 180 N + 2000
= (180 + 100) + 2000

$$= 20\ 000 \text{ litres}$$
$$= 20 \text{ m}^3$$

Internal dimensions of chambers

first chamber
$$\text{volume} = 20 \times \frac{2}{3} = 13.333 \text{ m}^3$$

length of chamber = twice the breadth, and assuming an average depth of 1.5 m,

$$\text{volume} = 2\,B \times B \times 1.5$$
$$= 3\,B^2$$

$$B = \sqrt{\frac{V}{3}}$$

$$= \sqrt{\frac{13.333}{3}}$$

$$= 2.108$$

length $= 2.108 \times 2 = 4.216$

internal dimensions
of first chamber $= 4.216 \times 2.108 \times 1.500$

internal dimensions
of second chamber $= 2.108 \times 2.108 \times 1.500$

Dimensions of biological filter

Assuming a circular filter bed on plan and an average depth of 1.8 m,

$$\text{volume} = 0.6 \times 100$$
$$= 60 \text{ m}^3$$

also $\text{volume} = \dfrac{\pi}{4}\,D^2 \times 1.8$

$\therefore$ diameter $= \sqrt{\dfrac{4\,V}{1.8\,\pi}}$

$$= \sqrt{\frac{4 \times 60}{1.8 \times 3.142}}$$

$$= 6.514$$

dimensions of
filter $= 6.514$ diameter, and 1.800 deep

Internal dimensions of humus tank

$$C = 30N + 1500$$
$$C = (30 \times 100) + 1.500$$
$$C = 4500 \text{ litres}$$
$$\text{volume} = 4.5 \text{ m}^3$$

Assuming a length of twice the breadth and an average depth of 1.5 m,

$$\text{volume} = 2\,B^2 \times 1.5$$

$$B = \sqrt{\frac{V}{3}}$$

$$= \sqrt{\frac{4.5}{3}}$$

$$= 1.225 \text{ m}$$

$$\text{dimensions} = 2.450 \times 1.225 \times 1.500$$

Questions (Use Chezy constants = 55 for all questions)

1. Calculate the velocity of flow in m/s and discharging capacity in m^3/s and litre/s of a 150 mm diameter drain flowing half-full and laid at a gradient of 1 in 120.

Answers: 0.9723 m/s; 0.008 594 m^3/s; 8.594 litre/s.

2. Calculate the velocity of flow in m/s and discharging capacity in m^3/s and litre/s of a 100 mm diameter drain flowing full when laid at a gradient of 1 in 50.

Answers: 1.23 m/s; 0.009 66 m^3/s; 9.66 litre/s.

3. Calculate the gradient required for a 150 mm diameter drain flowing half-full when the velocity required is 1 m/s.

Answer: 1 in 113.4

4. Calculate the gradient required for a 300 mm diameter surface-water sewer when flowing full at a velocity of flow of 0.8 m/s.

Answer: 1 in 354.5

5. Calculate the gradient required for a rectangular channel 300 mm wide when the depth of water is to be 150 mm flowing at a velocity of 1.5 m/s.

Answer: 1 in 101

6. Calculate the diameter and gradient for a surface water drain using the following factors:

   (a) rainfall intensity 50 mm/h;
   (b) area of surface 3000 m^2 ;
   (c) impermeability factor 0.9;
   (d) velocity of flow 1.2 m/s;
   (e) full-bore discharge.

Answers: 199.458 (use 200 mm diameter drain);
      gradient = 1 in 105.

7. A 150 mm diameter drain is to be laid at a gradient of 1 in 180. Calculate the velocity of flow when the drain is flowing half-full.

Answer: 0.7938 m/s (0.8 m/s approx.)

8. Define the following terms: terminal velocity; discharge unit; time of concentration; simultaneous demand factor.

9. Calculate the diameter of pipe and the gradient required to take the run-off from an area of one hectare of asphalt surface, using the following factors:
   (a) duration of storm 10 minutes;
   (b) impermeability factor 0.875;
   (c) time of concentration 20 minutes;
   (d) full-bore discharge;
   (e) velocity of flow 0.8 m/s.

Answers: 852 mm diameter; 1 in 1007 gradient.

10. By use of the discharge-unit method, find the diameter and gradient for a foul-water drain to take the discharges from 12 W.Cs (9-litre flush), 14 basins, and 6 sinks (assume public use).

Answers: 100 mm diameter; 1 in 45 gradient.

11. Calculate the diameter and gradient for a foul-water sewer to take the discharges from 50 houses. The average number of occupants is assumed to be four per house and the water consumption 225 litres per person, per day. The sewer is to be assumed to run half-full bore at a velocity of 0.8 m/s.

Answers: 125 mm diameter; 1 in 48 gradient.

12. By use of the discharge-unit method, find the diameter of a vertical stack to take the discharges from a three-storey building having the following fittings on each floor: 10 W.C.s (9-litre flush), 14 basins, and 2 sinks.

Answer: 125 mm diameter.

13. Calculate the internal dimensions of the septic tank, filter and humus tank to serve a population of 50 persons. Assume the following factors:
   (a) depth of septic tank 1.5 m;
   (b) depth of filter 1.8 m;
   (c) filter to be circular on plan;
   (d) depth of humus tank 1 m with length of twice the breadth.

Answers: (a) size of septic tank: first chamber = 3.126 × 1.563 × 1.500;
second chamber = 1.563 × 1.563 × 1.500;
   (b) size of filter 5.319 diameter, 1.800 depth;
   (d) size of humus tank 2.450 × 1.225 × 1.000.

Chapter 5

Rainwater pipes and gutters, flow over weirs

Rainwater pipes and gutters

The Building Research Establishment (BRE) has produced a design method for the sizing of rainwater pipes and gutters to match the rainfall intensity (*Digests* 188 and 189).

Rainfall intensity

Roof drainage calculations in the UK can usually be based on rainfall intensity of 75 mm/h, but short storms of a high intensity of 150 mm/h or more do occur and this high value should be used for situations where overflowing might cause serious damage.

It has been found that a storm producing a rainfall intensity of 75 mm/h may occur for 5 minutes once in 4 years, or for 20 minutes once in 50 years. An intensity of 150 mm/h may occur for 3 minutes once in 50 years, or for 4 minutes once in 100 years.

Flow load

The rate of run-off from a roof may be found from the following:

$$\text{rate of run-off} \atop (\text{m}^3/\text{s}) = \frac{\text{effective area (m}^2) \times \text{rainfall intensity (m/h)}}{3600}$$

Example 5.1. *Calculate the flow load from a roof having an effective area of 50 m² when the rainfall intensity is 75 mm/h.*

$$\text{rate of run-off} = \frac{50 \times 0.075}{3600}$$

$$= 0.001 \text{ m}^3 /\text{s (approx.)}$$
$$= 1 \text{ litre per second}$$

Allowance for wind (see Fig. 5.1)

In order to allow for the strength of the wind, it is suggested that the angle of descent of wind-driven rain should be taken as one unit horizontal for every two units of descent. This means that the effective area of the roof should be taken as the plan area plus half the elevation area. In Fig. 5.1 the effective area would be:

$$\text{effective area} = b + \frac{c}{2} \quad \text{(m)}$$
per m of eaves

One unit horizontal

Two units of descent

Angle of descent of wind driven rain

Effective length = $b + \frac{c}{2}$

b

$\frac{c}{2}$

c

θ θ

Fig.5.1 Effective length of roof

Run-off from vertical walls (see Fig. 5.2)

If a roof is bounded by vertical walls rising above the roof, some allowance should be made for the additional run-off from these walls. It is not safe to allow for absorption of water by the wall surfaces, even when these are of porous material.

If only one wall is involved, half its area should be added to the roof area to be drained. If the roof is contained within an angle formed by two or more walls, the direction from which these walls present the greatest projected area in elevation should be calculated and half that area added to the roof area to be drained calculated.

Eaves gutters

Each eaves gutter should be designed for the wind direction that would give the maximum rate of flow in the gutter. It is usually convenient in practice to calculate the flow load per metre run of eaves gutter.

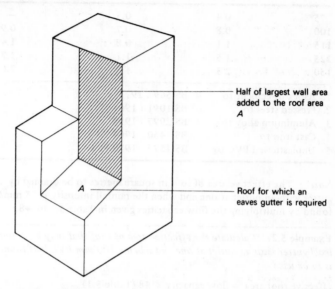

Half of largest wall area added to the roof area
A

Roof for which an eaves gutter is required

Fig. 5.2 Allowance for wall area

Effect of gutter angles

A square angle within 4 m of an outlet will reduce the flow capacity of an eaves gutter. This can be allowed for by multiplying the rate of run-off by one of the following factors:

1. angle within 2 m of the outlet
    sharp cornered × 1.2
    round cornered × 1.1
2. angle within 2–4 m of the outlet
    sharp cornered × 1.1
    round cornered × 1.05

Table 5.1 gives the flow capacities for level gutters, of half-round, segmental and ogee section, with outlet at one end. The flow capacities of gutters of other profiles and sizes may be found from the following formula:

$$\text{flow, litres per second} = \sqrt{\frac{A^3}{B}} \times 10^{-4}$$

where A = area of flow outlet (mm^2)
        B = width of water surface at outlet (mm)

Table 5.1

Nominal gutter size (mm)	True half-round gutter (1, 2, 5)	Nominal half-round segmental gutter (3, 4, 5)	Ogee gutter (2)	Ogee gutter (3, 4)
75	0.4	0.3	–	–
100	0.8	0.7	0.9	0.5
115	1.1	0.8	1.4	0.7
125	1.5	1.1	1.7	0.8
150	2.3	1.8	2.6	–

1. Asbestos-cement to   BS 569 : 1973
2. Pressed steel to   BS 1091 : 1963
3. Aluminium alloy to   BS 2997 : 1958
4. Cast iron to   BS 460 : 1964
5. Unplasticised PVC to   BS 4576 : 1970 Pt 1

Note: The effective area of roof in square metres to be drained by a level eaves gutter with an outlet at one end when the rainfall intensity is 75 mm/h, can be found by multiplying the flow capacities given in Table 5.1 by 48.

Example 5.2. *Calculate the effective area of roof that may be drained by a level gutter with an outlet at one end when a 100 mm PVC half-round gutter is to be used.*

effective roof area = flow capacity × 48 (Table 5.1)
= 0.8 × 48
= 38.4 m^2

For other rates of rainfall, the areas will be inversely proportional. Gutters fixed with a fall have a greater carrying capacity than when fixed level. Experiments by the Building Research Establishment have shown that an eaves gutter with a fall of 1 in 600 will carry up to 40 per cent more water than when fixed level. The increased flow also helps to prevent silting up of the gutter. Too great a fall looks unsightly and will sometimes allow water falling from the roof to miss the gutter. The rate of flow due to any fall on the gutter may be ignored and if the gutter is assumed to be fixed level the additional size will provide a safety margin.

Sizes of vertical rainwater pipes

The sizes of vertical rainwater pipes appropriate to the gutter sizes given in Table 5.1 are listed in Table 5.2. Figures are given for both sharp- and round-cornered outlets (see Fig. 5.3). Table 5.2 allows also for the position of the outlet in relation to the end of the gutter.

Example 5.3. *Calculate the sizes of eaves gutter and vertical rainwater pipes required to drain a roof 50 m long by 15 m (ridge to eaves) roof pitch 30°.*

Table 5.2 Minimum vertical rainwater pipe sizes (nominal diameter in mm) for various gutter sizes

Half-round gutter size	Sharp- or round-cornered outlet	Outlet at one end of gutter	Outlet not at end of gutter
75	sc	50	50
	rc	50	50
100	sc	63	63
	rc	50	50
115	sc	63	75
	rc	50	63
125	sc	75	89
	rc	63	75
150	sc	89	100
	rc	75	100

Sharp-cornered outlet

Water flowing in gutter

Round-cornered outlet

Fig.5.3 Types of gutter outlets

The plan length from ridge to eaves may be found by measurement from drawing or by calculation.

$$
\begin{aligned}
\text{by calculation, plan length} &= \cos 30^\circ \times 15 \\
&= 0.866 \times 15 \\
&= 12.99 \text{ m} \\
\text{vertical height} &= \tan 30^\circ \times 12.99 \\
&= 0.5774 \times 12.99 \\
&= 7.5 \text{ m} \\
\text{plan length + half the height} &= 12.99 + \frac{7.5}{2} \\
&= 16.74 \text{ m} \\
\text{effective area of roof per m of eaves} &= 16.74 \text{ m}^2
\end{aligned}
$$

With a rainfall intensity of 75 mm/h the rate of run-off per m of eaves would be:

$$\text{rate of run-off} = \frac{16.74 \times 0.075}{3600}$$

$$= 0.000\,35 \text{ m}^3/\text{s per m}$$
$$= 0.35 \text{ litre/s per m}$$

From Table 5.1, a 150 mm (nominal) true half-round gutter could cope with a flow from a length of eaves of

$$\frac{2.3}{0.35} = 6.57 \text{ m}$$

Since the gutter outlet receives the flow of water from either side (see Fig. 5.3) and the outlet takes water from only half of the length of gutter to the next outlet, outlets would be required at not more than double the above distance apart.

distance between outlets $= 2 \times 6.57 = 13.14$ m

Figure 5.4 shows a possible solution to the spacing of the rainwater pipes based on this distance.

Fig.5.4 Spacing of rainwater pipes

Diameter of pipes

Assuming the use of round-cornered outlets, Table 5.2 shows that 100 mm diameter pipes would be suitable for a 150 mm half-round gutter with the outlets not at the end of the gutter.

Flat roofs

Although the term 'flat roofs' is commonly used, these have a fall of about 1 in 60.

Example 5.4 (see Fig. 5.5) *A flat roof measuring 40 m by 30 m is to be drained by rectangular gutters down each long side. Each gutter is to be provided with an outlet at each end. If a rainfall intensity of 75 mm/h is to be assumed, calculate the size of gutters and the diameter of the outlets and rainwater pipes.*

70

Rectangular gutter 300 mm x 130 mm deep

Fall 1 in 60

30.000

Outlet

40.000

Outlet

Fig.5.5 Flat-roof drainage

Since each gutter will have a high point at the centre of the gutter, each outlet will take the discharge from one-quarter of the roof, and the gutter will also have to be designed to take this discharge. Therefore,

$$q = \frac{\text{roof area m}^2 \times \text{rainfall intensity m} \times 1000}{4 \times 3600} \text{ litre/s}$$

$$q = \frac{40 \times 30 \times 0\,075 \times 1000}{4 \times 3600} \text{ litre/s}$$

$$= 6.25 \text{ litre/s}$$

The Building Research Establishment *Digest* 189, part 2, gives a formula for the flow capacity of a gutter discharging freely, and this is approximately equal to:

$$q = \sqrt{\frac{A^3}{B}} \times 10^{-4} \text{ litre/s}$$

$$\left[\frac{q}{10^{-4}}\right]^2 = \frac{A^3}{B}$$

$$\left[\frac{q}{10^{-4}}\right]^2 \times B = A^3$$

$$A = \sqrt[3]{\left[\frac{q}{10^{-4}}\right]^2 \times B}$$

where A =   area of water flow at outlet (mm^2)
          B =   width of water surface at outlet (mm)

71

Assume a gutter width of 300 mm. To avoid swirl, the width should be approximately equal to twice the diameter of the outlet (which can be checked later).

$$A = \sqrt[3]{\left[\frac{6.25}{0.0001}\right]^2 \times 300}$$

$$= \sqrt[3]{62\,500^2 \times 300}$$

$$= 10\,520 \text{ mm}^2$$

$$\text{depth of water flow at outlet} = \frac{A}{B} = \frac{10\,520}{300}$$

$$= 35 \text{ mm}$$

For free discharge, the maximum depth of flow in the gutter is twice the depth at the outlet, therefore the maximum flow depth is $2 \times 35 = 70$ mm. For large gutters, however, a freeboard or height above the water level at the top of the gutter when the water is flowing at its maximum rated depth should be 50 to 60 mm; therefore the final depth of gutter is $70 + 60 = 130$ mm.

$\therefore$ size of rectangular gutter = 300 mm $\times$ 130 mm deep.

Rainwater pipes

Where a gutter discharges directly into a rainwater pipe, the rainwater pipe (or gutter outlet) should be designed to allow free discharge of water from the gutter.

When sizing the outlet to a rainwater pipe leading directly from a gutter, the design head H_d should be taken as the depth of flow at the gutter outlet H_o in Fig. 5.6. If, however, the gutter discharges into a box receiver to which the rainwater pipe connects, the head is the depth of water in the box (see Fig. 5.7).

Fig.5.6 Depth of flow from gutter leading directly into a rainwater pipe

The diameter of the rainwater pipe can be reduced to two-thirds of the effective outlet diameter, provided that the transition is gradual over a length of not less than the diameter of the outlet (see Fig. 5.8).

Fig.5.7 Depth of flow from gutter leading to a box receiver

Fig.5.8 Type of inlet to allow a reduction in a rainwater pipe diameter

If the outlet is covered with a grating, the effective discharge capacity of the outlet must be calculated from the unobstructive area of the grating. The discharge capacity of a rainwater pipe depends upon the type of outlet, the smoothness of the pipe bore, the diameter of the outlet, the depth of water over the outlet, and whether the water swirls and forms a vortex about the outlet.

Swirl can be neglected if the centre line of the outlet is at a distance less than its diameter from the nearest vertical side of the gutter or outlet box (see Fig. 5.9). Swirl can also be suppressed by placing a vertical guide vane along the axis of the gutter, over the rainwater pipe outlet.

In the absence of swirl the outlet acts as a weir for values of the ratio H_1/D up to one-third and the discharge capacity of outlet in litres per second is:

$$q = \frac{D}{5000} \sqrt{H_1^3}$$

Fig. 5.9 Dimensions of gutter or box to avoid swirl at the outlet

where D = diameter of outlet (mm)
         H_1 = head of water over outlet (mm)

Where H_1/D is greater than $\frac{1}{3}$, the outlet acts as an orifice and the discharge capacity is:

$$q = \frac{D^2}{15\,000} \sqrt{H_1} \text{ litre/s}$$

Swirl will occur when the centre of the outlet is at a distance greater than its diameter from the nearest vertical side of the gutter or box (unless suppressed by a vane). The outlet then acts as a weir with the same discharge rate as in the absence of swirl for values of H_1/D up to $\frac{1}{4}$; above this, the outlet acts as an orifice but swirling will reduce its discharging capacity to:

$$q = \frac{D^2}{20\,000} \sqrt{H_1} \text{ litre/s}$$

If it is assumed that swirl will be absent, the first formula may be used to find the diameter of the rainwater pipe in example 5.4 and a check made to find if the ratio of H_1/D is less than $\frac{1}{3}$.

$$q = \frac{D}{5000} \sqrt{H_1^3} \text{ litre/s}$$

by transposition

$$\frac{q}{\sqrt{H_1^3}} = \frac{D}{5000}$$

$$\therefore \quad D = \frac{q \times 5000}{\sqrt{H_1^3}}$$

by substitution

$$D = \frac{6.25 \times 5000}{\sqrt{35^3}}$$

$$= \frac{31\,250}{\sqrt{42\,875}}$$

$$= \frac{31\,250}{207}$$

$$= 150.97 \text{ mm}$$

$$= 151 \text{ mm (approx.)}$$

If the effective outlet diameter is reduced gradually over a length of not less than the diameter of the outlet, the diameter of the rainwater pipe may be reduced by two-thirds.

diameter of rainwater pipe = $150.97 \times \frac{2}{3}$ = 100.64 mm

check on the ratio $H_1/D = \frac{35}{150.97} = 0.232$.

This is less than $\frac{1}{3}$, and therefore the first formula would be satisfactory. Therefore 4–100 mm diameter pipes would be satisfactory. Some designers may choose 4–125 mm diameter pipes to allow for exceptional storms.

The Institute of Plumbing provide a formula for the sizing of vertical rainwater pipes in the form of:

$$RA = \frac{3.2\,D^{5/2}}{1000}$$

or

$$RA = \frac{3.2\,\sqrt{D^5}}{1000}$$

where

RA = roof area in m^2 for a rainfall intensity of 75 mm/h
D = diameter of rainwater pipe in mm

by transposition

$$D = \sqrt[5]{\left[\frac{RA \times 1000}{3.2}\right]^2}$$

roof area

$$= \frac{40 \times 30}{4} = 300 \text{ m}^2$$

$$\therefore \quad D = \sqrt[5]{\left[\frac{300 \times 1000}{3.2}\right]^2}$$

$$= \sqrt[5]{93\,750^2}$$

75

$$= 97.43 \text{ mm}$$

By use of this formula, it is seen that 4–100 mm diameter pipes would be satisfactory.

The Institute of Plumbing also provide a formula for the sizing of the inlet to the rainwater pipe in the form of:

$$RA = \frac{9}{1000} \, db^{3/2} \text{ or}$$

$$\frac{9}{1000} \, d \, \sqrt{b^3}$$

where
RA = roof area in m^2 for a rainfall intensity of 75 mm/h
d = diameter of outlet in mm
b = head of water over inlet in mm

by transposition

$$d = \frac{300\,000}{\sqrt{b^3} \times 9}$$

$$= \frac{300\,000}{\sqrt{35^3} \times 9}$$

$$= \frac{300\,000}{207 \times 9}$$

$$= 161 \text{ mm} \qquad \text{(BRE 151 mm)}$$

This gives a 10 mm increase in diameter over the BRE method, which is negligible.

Check on gutter width: To avoid swirl at the outlet and a vortex being formed, the width of the gutter should be less than twice the diameter of the outlet: 2 × 151 = 302 mm. A 300 mm width is therefore suitable.

Flow of water over weirs

When the sides of an orifice are produced, so that they extend beyond the free surface of the water, it is called a weir. In other words, water flows through an orifice but it flows over a weir. Small weirs are called notches, which usually take the form of thin vertical plates with accurately shaped openings, either rectangular or triangular. and gauging the flow of water.

For example, if the flow of water in a small stream or channel is required, a dam is constructed across the stream and the water allowed to pass through a notch cut in a board or metal plate. The water flowing over a weir or notch is called 'nappe', which corresponds to the jet leaving an orifice.

Ventilation openings

Water flowing over a weir or notch must never be allowed to cling to the downstream face of the plate, but must spring clear. In order to achieve this, ventilation openings must be made in the plate as shown in Fig. 5.10.

Fig.5.10 Ventilation openings

In the absence of ventilation openings, a pocket of air will be trapped under the nappe and this air will soon be absorbed by the falling water. This will cause a partial vacuum under the nappe which will allow the water to be forced back by the atmospheric pressure acting upon the free water surface until it clings to the downstream face of the weir plate.

Head over crest (H) (see Fig. 5.11)

The water level in the upstream channel falls as the stream approaches the weir plate, so that the head over crest is less than the true head which should be measured in still water. The true head, therefore, should be located at a point some distance back from the upstream face of the weir plate. A measuring gauge for this purpose is usually placed upstream at a minimum distance of six times the head over crest.

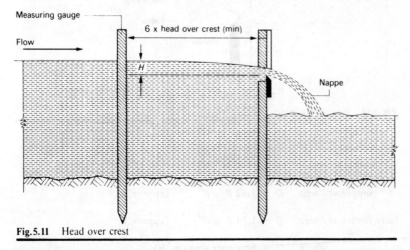

Fig.5.11 Head over crest

Rectangular weirs (see Fig. 5.12)

The edges of the openings may or may not be bevelled. If they are bevelled so as to form a sharp-edged weir, the form of nappe will be different from that formed when the edges are square.

If the sides of the weir are cut square but the edge is sharp, the nappe flows over at full width but the depth will show a contraction due to the sharp-edged crest. This type of weir is called a weir with suppressed end contractions or a suppressed rectangular weir.

If the ends of openings as well as the crest are sharp-edged, the end contractions increase with increasing head. The width of the nappe, however, decreases due to the end contractions. This type of weir is called a fully contracted rectangular weir.

Fig.5.12   Rectangular weirs

Calculations

The formulae used for weirs are as follows:

$$\text{suppressed weir} \quad Q = 1.82 \, B \, \sqrt{H^3} \quad \text{(approx.)}$$

$$\text{fully contracted weir} \quad Q = 1.74 \, B \, \sqrt{H^3} \quad \text{(approx.)}$$

where Q = flow over weir (m³/s)
        B = breadth of weir (m)
        H = head over weir (m)

Example 5.5. *A suppressed rectangular weir is to be used to measure the flow of water in a stream. Calculate the rate of flow in litres per second when the breadth is 600 mm and the head of water over the crest is 150 mm.*

$$Q = 1.82 \ B \ \sqrt{H^3}$$

$$= 1.82 \times 0.6 \times \sqrt{0.15^3}$$

$$= 0.063 \ 34 \ m^3 /s$$

$$= 63.34 \ litre/s \qquad (approx.)$$

Example 5.6. *Calculate the rate of flow over a fully contracted rectangular weir having the same dimensions given in example 5.5.*

$$Q = 1.74 \ B \ \sqrt{H^3}$$

$$= 1.74 \times 0.6 \times \sqrt{0.15^3}$$

$$= 0.0606 \ m^3 /s$$

$$= 60.60 \ litre/s \qquad (approx.)$$

Triangular weir (see Fig. 5.13)

For smaller flows of water a triangular weir may be used instead of a rectangular weir. The weir is vee-shaped with the angle of the vee equal to θ. The head of water over the weir is the true head described previously.

Elevation   Section

Fig.5.13 Triangular weir

Calculations

The formula used when the angle of vee is 90° is given by:

$$Q = 1.4 \ \sqrt{H^5} \qquad (approx.)$$

For an angle other than 90°, the formula becomes:

$$Q = 1.4 \tan\frac{\theta}{2} \ \sqrt{H^5} \qquad (approx.)$$

where
$$Q = \text{flow over weir } (m^3 /s)$$
$$H = \text{head over weir } (m)$$

Example 5.7. *Calculate the rate of flow in litres per second over a 90° triangular weir when the head over the notch is 100 mm.*

$$Q = 1.4 \sqrt{H^5}$$

$$= 1.4 \sqrt{0.1^5}$$

$$= 0.004\ 43\ \text{m}^3/\text{s}$$

$$= 4.43\ \text{litre/s} \qquad (\text{approx.})$$

Example 5.8. *Calculate the rate of flow over a triangular weir having the same head as that given in example 5.7 but with an angle of 45°.*

$$Q = 1.4 \tan\frac{\theta}{2} \sqrt{H^5}$$

$$= 1.4 \times \frac{1}{2} \sqrt{0.1^5}$$

$$= 1.4 \times 0.5 \sqrt{0.1^5}$$

$$= 1.4 \times 0.5 \times 0.003\ 162$$

$$= 0.0022\ \text{m}^3/\text{s}$$

$$= 2.2\ \text{litre/s} \qquad (\text{approx.})$$

Questions

1. Calculate the flow load in litres per second from a roof having an effective area of 80 m^2 when the rainfall intensity is 75 mm/h.

Answer: 1.67 litre/s

2. Calculate the effective area of roof that may be drained by a level gutter with an outlet at one end when a 150 mm PVC half-round gutter is to be used.

Answer: 110.4 m^2

3. Calculate the sizes of an eaves gutter and rainwater pipes to drain a roof 30 m long when the plan length from ridge to eaves is 10 m and the vertical height 8 m. Draw a plan of the gutter to show the positions of the vertical rainwater pipes.

Answers:

(a)

(b)

Fig.5.14 Alternative answers to question 3, Chapter 5

4. A flat roof 50 m by 20 m is to be drained by rectangular gutters down each long side. Each gutter is to be provided with an outlet at each end. If a rainfall intensity of 75 mm/h is to be assumed, calculate the sizes of the gutters and the diameter of the rainwater pipes. (Allow freeboard depth of 60 mm.)

Answers: gutter size = 250 mm × 130 mm deep
diameter of pipes = 4–100 mm
outlet diameter = 150 mm

5. A flat roof measuring 20 m by 15 m is to be laid to falls to one rainwater pipe. Assuming a rainfall intensity of 75 mm/h, calculate the diameter of the pipe.

Answer: 100 mm diameter

6. If a gutter discharges from a roof into a box receiver, calculate the minimum length of the box when the box is at the end of the gutter. Height of fall of water 150 mm and depth of water at the discharge point 30 mm.

Answer: 134.2 mm

7. Sketch the type of roof outlet to a rainwater pipe that will provide a flow characteristic sufficient to reduce the pipe diameter by two-thirds. Describe how swirl may be prevented at a roof outlet.

8. Describe the methods used to find the rate of flow in a small stream or channel.

9. A fully contracted rectangular weir has a crest width of 1.200 and a head of water over the crest of 200 mm. Calculate the rate of flow over the weir in litres per second.

Answer: 186.756 litre/s

10. A triangular weir has a vee-angle of 60° and a head of water over the weir of 150 mm. Calculate the rate of flow over the weir in litres per second.

Answer: 10.565 litre/s

Chapter 6

Electrical and gas installations

Electrical installations

Power

The unit of power is the watt (W). If the potential difference between two points of a conductor is 1 V, when carrying a current of 1 A, the power dissipated between the two points is equal to 1 W.

$$W = V I \qquad\qquad [6.1]$$

where V = potential difference in volts
I = current in amperes

Example 6.1. *An electric cable will carry a current of 10 A when the supply is 240 V. Calculate the power load that may be carried by the cable.*

$$
\begin{aligned}
W &= V I \\
&= 240 \times 10 \\
&= 2400 \text{ W} \\
&= 2.4 \text{ kW}
\end{aligned}
$$

Example 6.2. *A 3 kW electric water heater is connected to a 240 V supply. Calculate the current flowing in the cable when the heater is in use.*

$$
\begin{aligned}
W &= V I \\
I &= \frac{W}{V}
\end{aligned}
$$

$$= \frac{3000}{240}$$

$$= 12.5 \text{ A}$$

Fuses or miniature circuit-breakers

In order to prevent overloading of the cable and possible overheating of the wires, a fuse or miniature circuit-breaker is required which is connected between the live conductor.

In example 6.2 the rating of the fuse or miniature circuit-breaker would normally be 13 A.

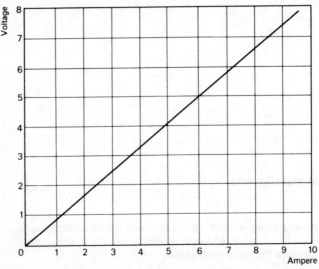

Fig.6.1 Graphical representation of Ohm's law. The gradient gives the resistance ie $R = \frac{V}{A}$

Ohm's law (See Fig. 6.1)

The current passing through a conductor at constant temperature is proportional to the potential difference between its ends.

$$\text{resistance} = \frac{\text{potential difference}}{\text{current}}$$

$$\text{or } R = \frac{V}{I} \text{ ohms}$$

We can also write

$$V = I R \qquad\qquad [6.2]$$

Example 6.3. *The potential difference across a resistance through which a current of 30 A is flowing is 50 V. Calculate the resistance.*

$$R = \frac{V}{I}$$

$$= \frac{50}{30}$$

$$= 1.667 \ \Omega$$

Example 6.4. *What current is flowing through a conductor having a resistance of 50 ohms when the potential difference across the conductor is 240 V?*

$$R = \frac{V}{I}$$

$$I = \frac{V}{R}$$

$$= \frac{240}{50}$$

$$= 4.8 \ A$$

Heating effect of current

It is possible to combine the formulas for power and resistance. The formulas [6.1] and [6.2] may be combined as follows:

$$W = V I \tag{6.1}$$
$$V = I R \tag{6.2}$$

Substituting [6.2] in [6.1],

$$W = I R \times I$$
$$W = I^2 \times R \tag{6.3}$$

Also from [6.2],

$$I = \frac{V}{R}$$

and substituting in [6.1]

$$W = V \times \frac{V}{R}$$

$$W = \frac{V^2}{R} \tag{6.4}$$

Note: From formulas [6.3] and [6.4] it can be seen that the power varies either with the square of the current, or the square of the voltage.

Example 6.5. *Calculate the current flowing in and the resistance of the filament of a 150 W tungsten filament lamp (when in use) when connected to a 240 V supply.*

85

$$W = V I$$

$$I = \frac{W}{V}$$

$$= \frac{150}{240}$$

$$= 0.625$$

by Ohm's law $V = I R$

$$R = \frac{V}{I} = \frac{240}{0.625}$$

resistance $= 384 \ \Omega$

Check by using formula [6.3] :

$$W = I^2 \times R$$
$$= 0.625^2 \times 384$$
$$= 150$$

Example 6.6. *Calculate the maximum current which can be taken and the resistance of an electric heater element rated at 240 V and 1500 W.*

$$W = V I$$

$$I = \frac{W}{V}$$

$$= \frac{1500}{240}$$

$$= 6.25$$

$$R = \frac{V}{I}$$

$$= \frac{240}{6.25}$$

$$= 38.4 \ \Omega$$

Losses at high voltages

Electricity is transmitted in the National Grid system at high voltages of 132, 275, or 400 kV. The power wasted as heat in these high-voltage transmission cables is very low when compared with the power wasted as heat in low-voltage transmission cables.

Example 6.7. *Calculate the power wasted as internal energy in a cable when 15 kW is transmitted through a cable having a resistance of 0.8 Ω: (a) at 240 V; (b) 275 000 V.*

(a) $\qquad I = \dfrac{W}{V}$

$$I = \frac{15\ 000}{240}$$

$$= 62.5$$

Power loss in cable (low voltage)

$$W = I^2\ R$$
$$= 62.5^2 \times 0.8$$
$$= 3\ 125 \text{ or } 3.125 \text{ kW}$$

(b) $\qquad I = \dfrac{W}{V}$

$$= \frac{15\ 000}{275\ 000}$$

$$= 0.0545$$

Power loss in cable (high voltage)

$$W = I^2\ R$$
$$= 0.0545^2 \times 0.8$$
$$= 0.002\ 38 \text{ or } 2.38 \text{ mW}$$

We see that the power wasted at high voltages is very low indeed, or almost negligible.

Note: One 400 kV line has three times the power-carrying capacity of one 275 kV line and eighteen times the carrying capacity of one 132 kV line.

Transformers (See Fig. 6.2)

These consist of two distinct windings or coils wound on a laminated iron core. The winding connected on the a.c. supply is known as the primary winding and the secondary winding is connected to the load. An alternating current flows through the primary winding, causing a resultant magnetic flux in the iron core. When this magnetic flux cuts the secondary winding, there is an electromotive force induced in the winding which sends an alternating current through the load.

For a transformer the following relationships apply, if losses are neglected:

$$\frac{\text{primary voltage}}{\text{secondary voltage}} = \frac{\text{primary turns}}{\text{secondary turns}} = \frac{\text{secondary current}}{\text{primary current}}$$

$$\frac{V_1}{V_2} = \frac{N_1}{N_2} = \frac{I_2}{I_1}$$

where
$\quad V_1$ = primary voltage
$\quad V_2$ = secondary voltage
$\quad N_1$ = number of primary turns
$\quad N_2$ = number of secondary turns
$\quad I_1$ = primary current
$\quad I_2$ = secondary current

Laminated core to reduce magnitude of eddy currents

A.C. supply

Primary winding with N_1 turns

Input
V_1 and A_1

Output
V_2 and A_2

Secondary winding with N_2 turns

Fig.6.2 Transformer

Example 6.8. *A transformer has 1500 turns in the primary winding and 300 turns in the secondary winding. If the secondary voltage and current are 240 V and 15 A respectively, calculate:* (a) *the primary current;* (b) *the primary voltage.*

(a)
$$\frac{N_1}{N_2} = \frac{I_2}{I_1}$$

$$N_1 \times I_1 = N_2 \times I_2$$

$$I_1 = \frac{N_2 \times I_2}{N_1}$$

$$= \frac{300 \times 15}{1500} = 3$$

primary current = 3 amperes

(b)
$$\frac{V_1}{V_2} = \frac{N_1}{N_2}$$

$$V_1 \times N_2 = V_2 \times N_1$$

$$V_1 = \frac{V_2 \times N_1}{N_2}$$

$$= \frac{240 \times 1500}{300} = 1200$$

primary voltage = 1200

It will be noticed that the primary high voltage has a lower current than the lower secondary voltage. This results in a lower power loss in the primary circuit as shown in example 6.7.

Area of conductor (see Fig. 6.3)

The resistance offered by a conductor is inversely proportional to its cross-sectional area. In other words, if the cross-sectional area of a conductor is doubled, then resistance is halved, or vice versa.

Four 2 mm diameter wires with equal area of one 4 mm diameter wire

Fig 6.3 Cross-sectional areas of wires

The areas of circles are to each other as the squares of their diameters, or if a number of small wires are to have the same cross-sectional area as one large wire,

$$N = \frac{D^2}{d^2}$$

where
N = number of small-diameter wires
D = diameter of large wire
d = diameter of small wire

Example 6.9. *A certain wire has a diameter of 4 mm and a resistance of 0.5 Ω. What will be the resistance of a 2 mm diameter wire carrying the same values as the 4 mm diameter?*

The number of smaller diameter wires of equal area of the larger diameter wire is given by

$$N = \frac{D^2}{d^2} = \frac{4^2}{2^2} = \frac{16}{4} = 4$$

The large-diameter wire, therefore, is four times greater in area than the small-diameter wire. Therefore, the small-diameter wire will offer four times the resistance of the large-diameter wire.

$4 \times 0.5 = 2 \ \Omega$

Length of conductor

If all other values are equal, the resistance of a conductor is proportional to its length.

Example 6.10. *A wire 2 km long has a resistance of 100 Ω. What is the resistance of a piece of this wire 100 m long?*

2000 m offers a resistance of 100 Ω

$\therefore$ 100 m offers a resistance of

$$\frac{100}{2000} \times \frac{100}{1} = 5 \ \Omega$$

Effect of temperature

With most conductors, a rise in temperature results in an increase in resistance. The resistance of some substances, including carbon, becomes smaller instead of greater as the temperature rises. Certain alloys are also almost unaffected by a change in temperature.

Note: One type of fire-control system operates on the effect of the increase of resistance due to temperature rise. In the system, two copper wires are fixed at high level. One of the wires is insulated against heat and the other left uninsulated. A small electric current is passed through both wires and if a fire occurs, the heat causes a reduction in the flow of current through the uninsulated wire due to the temperature rise. The difference in current flowing through the two wires causes an electric-magnetic valve to open and water is allowed to flow through pipework to open sprinklers.

Root mean square value (r.m.s.) (See Fig. 6.4)

Alternating current and voltage values fluctuate between zero and maximum potential, while direct current values remain steady. In order to obtain the same values for a.c. as are present in d.c., the r.m.s. value is used. The r.m.s. value of a current of 1 A a.c. gives the same heating effect as 1 A d.c.; and similarly for a.c. voltage.

The relationship of r.m.s. value to maximum value a.c. is:

$$\frac{\text{r.m.s. value}}{\text{maximum value}} = 0.707$$

Equivalently,

$$\text{maximum value} = \text{r.m.s. value} \times 1.414$$

Example 6.11.

(a) *Calculate the maximum value of a.c. voltage when the r.m.s. value is 240 V.*

(b) *Calculate the r.m.s. value of an a.c. current which has a maximum value of 30 A.*

(a) maximum value = r.m.s. value × 1.414
    = 240 × 1.414
    = 339.36 V

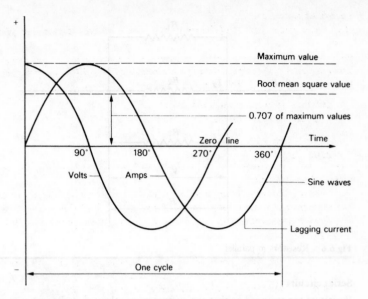

Fig.6.4 Root mean square value

(b)   r.m.s. value = maximum value × 0.707
                    = 30 × 0.707
                    = 21.21 A

Series and parallel circuits

Graphical representation of resistor

Fig.6.5 Resistors in series

Fig.6.6   Resistors in parallel

Series circuits

If circuits are arranged so that the current passes through resistors consecutively, as shown in Fig. 6.5, the resistors are said to be connected in series.

The same current must pass through all the resistors. If R_1, R_2, R_3, etc. are the resistors, then:

$$V = I R$$
$$V = I R_1 + I R_2 + I R_3$$
$$I R = I R_1 + I R_2 + I R_3$$

The total resistance is therefore the sum of the individual resistances. Dividing throughout by the current,

$$R = R_1 + R_2 + R_3$$

Example 6.12.   *What is the total resistance of three resistors connected in series when the individual resistances are 10 Ω, 5 Ω and 3 Ω?*

total resistance = 10 + 5 + 3 = 18 Ω

Parallel circuits

If circuits are arranged side by side with their corresponding ends joined together, as shown in Fig. 6.6, the resistors are said to be connected in parallel.

The same potential difference will apply to each resistor but they will share the same current. Current I will be divided into I_1, I_2 and I_3 through resistances R_1, R_2 and R_3.

$$\text{total current} \quad I = I_1 + I_2 + I_3$$
$$= \frac{V}{R_1} + \frac{V}{R_2} + \frac{V}{R_3}$$
$$\frac{V}{R} = \frac{V}{R_1} + \frac{V}{R_2} + \frac{V}{R_3}$$

Dividing throughout by V,

$$\frac{1}{R} = \frac{1}{R_1} + \frac{1}{R_2} + \frac{1}{R_3}$$

Example 6.13. *Two resistors of 5 and 20 ohms are connected in parallel when the potential difference is 25 volts. Calculate: (a) the total resistance; (b) the total current required; (c) the current through each resistor.*

(a)

$$\frac{1}{R} = \frac{1}{R_1} + \frac{1}{R_2}$$

$$= \frac{1}{5} + \frac{1}{20}$$

$$= \frac{4 + 1}{20}$$

$$= \frac{5}{20}$$

$$\therefore \quad R = \frac{20}{5}$$

$$= 4 \ \Omega$$

(b)

$$V = I R$$

$$I = \frac{V}{R}$$

$$= \frac{25}{4}$$

$$= 6.25 \ \text{A}$$

(c)

$$V = I_1 R_1$$

$$I_1 = \frac{V}{R_1} = \frac{25}{5} = 5 \ \text{A}$$

$$I_2 = \frac{V}{R_2} = \frac{25}{20} = 1.25 \ \text{A}$$

Note: The total current is the same as given in part (b).

Power factor

The term 'power factor' applies to alternating current only and is the proportion of current which can be used as energy. A power factor of 0.8 means that 80 per cent of the current supplied is used as energy and the remaining 20 per cent is idle.

In an inductive circuit, such as produced by a motor or solenoid, the electromagnetic force opposes the applied voltage which causes the current wave to lag behind the voltage wave. This results in loss of power due to the voltage and current being out of step (see Fig. 6.4).

$$\text{power factor} = \frac{\text{true power}}{\text{apparent power}}$$

True power is measured in watts or kilowatts and apparent power is the product of volts and amperes.

Kilovolt-ampere, kVA

This is the measurement of apparent power in an a.c. supply.

Example 6.14. *Find the apparent power in a 1000 V and 100 A a.c. supply.*

1000 V × 100 A = 1 kV × 100 A = 100 kVA

Example 6.15. *A diesel-electric generator is rated at 500 kVA. If the power factor for the machine is 0.85, find the true power that may be obtained.*

$$
\begin{aligned}
\text{kW} &= \text{kVA} \times \text{power factor} \\
\text{power} &= 500 \times 0.85 \\
&= 425 \text{ kW}
\end{aligned}
$$

Example 6.16. *A 2250 W motor on a 240 V single-phase supply is found to take 15 A and 3000 W on full load. Calculate the power factor and percentage efficiency of the motor.*

$$\text{power factor} = \frac{W}{V\,A} = \frac{3000}{240 \times 15} = 0.833$$

$$\text{efficiency} = \frac{\text{power output}}{\text{power input}} \times \frac{100}{1}$$

$$= \frac{2250}{3000} \times \frac{100}{1} = 75 \text{ per cent}$$

Line and phase voltages

In a four-wire, three-phase supply there are three line conductors carrying a live voltage of 240 each and one neutral return conductor. The voltage between any one of the line conductors and the neutral will provide a single-phase voltage of 240. The voltage between any two line conductors will give a line voltage of 1.73 times the phase voltage.

$$\text{line voltage} = \text{phase voltage} \times \sqrt{3}$$

$$= \text{phase voltage} \times 1.73$$

and

$$\text{phase voltage} = \frac{\text{line voltage}}{1.73}$$

If the phase voltage is 240 the line voltage will be:

$$\text{line voltage} = 240 \times 1.73$$

$$= 415.2 \text{ (usually given as 415 V)}$$

94

Secondary windings of transformer
with star connection

Line 1

240 V

415 V

Neutral

240 V

415 V

Line 2

240 V

415 V

Line 3

Earth

Line voltage = 415
Phase voltage = 240

Fig.6.7 Line and phase voltage

Three-phase supplies

For buildings other than dwellings and other small buildings, a three-phase supply is required to meet the higher power load. Electricity boards usually require a three-phase supply when a motor requires 3.73 kW of power or over. It can be shown that three-phase supply is more efficient than single-phase.

Example 6.17. *Calculate the gain in power, expressed as a percentage, of three-phase over single-phase supply, assuming unity power factor and with 240 V and 10 A respectively.*

$$\text{single-phase power} = VI$$
$$= 240 \times 10$$
$$= 2400 \text{ W}$$

$$\text{three-phase power} = VI \sqrt{3}$$

$$= 240 \times 10 \times 1.73$$
$$= 4152 \text{ W}$$
$$\text{gain in power of three-phase} = 4152 - 2400$$
$$= 1752 \text{ W}$$

95

gain in power of three-phase,
over single-phase expressed $= \dfrac{1752}{2400} \times \dfrac{100}{1}$
as a percentage

$= 73$ per cent

Therefore for 50 per cent more wire, a three-phase supply provides 73 per cent more power for the same values of voltage and current.

Water heating

It has been found by experiment that the heat energy or the specific heat capacity of water (s.h.c.) is approximately equal to 4.2 kJ/kg K. In other words, it requires 4.2 kJ of energy to raise the temperature of 1 kg mass of water through 1 °C.

Example 6.18. *Calculate the amount of energy required to raise the temperature of 80 kg of water through 60 °C assuming no heat losses.*

$$\text{energy} = \text{s.h.c.} \times \text{kg} \times \text{K}$$
$$= 4.2 \times 80 \times 60$$
$$\text{energy required} = 20\,160 \text{ joules}$$

$$\text{Since kilowatts} = \frac{\text{kilojoules}}{\text{seconds}}$$

the power in kilowatts required, if the water is to be heated in one hour, is:

$$\text{power} = \frac{20\,160}{60 \times 60}$$

$$= 5.6 \text{ kW}$$

In practice, an allowance will have to be made for the loss of heat from the hot-water storage vessel, even though the vessel may be covered with a good thickness of insulation, say 76 mm. It is not usual to heat stored water by electricity in only one hour. For stored water therefore, we may write:

$$\text{power} = \frac{\text{s.h.c.} \times \text{temperature rise (K)} \times \text{mass of water (kg)} \times 100}{\text{heating time in seconds} \times \text{efficiency (per cent)}} \text{ kW}$$

Note: If the heat losses are 20 per cent the efficiency will be 80 per cent.

Kilowatt hour

The unit of electricity is 1 kilowatt of electrical power used for 1 hour, or 1 kWh

Example 6.19. *Find the number of kWhs and the cost when a 3 kW fire is used for 5 hours and the cost of 1 kWh is 5p.*

unit $= 3 \text{ kW} \times 5 \text{ h} = 15 \text{ kWh}$
cost $= 15 \times 7p = 105p$

Energy, work: The base unit of energy or work is the joule, which is equal to 1 N m.

$$\text{since } 1 \text{ watt} = \frac{1 \text{ N m}}{\text{s}}$$

$$1 \text{ watt} = \frac{1 \text{ joule}}{\text{s}}$$

or joules = watts × seconds

Therefore, one unit (kWh) of electricity is given by:

$$
\begin{aligned}
\text{energy} &= 1000 \times 60 \times 60 \\
&= 1000 \times 3600 \\
&= 3\ 600\ 000 \text{ joules} \\
&= 3.6 \times 10^6 \text{ joules or } 3.6 \text{ MJ}
\end{aligned}
$$

Example 6.20. *Calculate the power in kW required to raise the temperature of 45 kg of water from 15 °C to 70 °C in 2 hours when the heat losses are 20 per cent.*

$$\text{power} = \frac{4.2 \times (70 - 15) \times 45 \times 100}{2 \times 3600 \times 80} \text{ kW}$$

$$= 1.8 \text{ kW}$$

A 2 kW electric immersion heater would be required.

Example 6.21. *Calculate the time in hours that it will take to raise the temperature of 136 kg of water from 10 °C to 60 °C by means of a 3 kW immersion heater when the heat losses are 10 per cent.*

$$\text{power} = \frac{\text{s.h.c.} \times \text{kg} \times \text{temperature rise (K)} \times 100}{\text{heating time in seconds} \times \text{efficiency}} \text{ kW}$$

By transposition,

$$
\begin{aligned}
\text{time in seconds} &= \frac{\text{s.h.c.} \times \text{kg} \times \text{K} \times 100}{\text{kW} \times \text{efficiency}} \\
&= \frac{4.2 \times 136 \times 50 \times 100}{3 \times 90} \\
&= 10\ 577.777 \\
\text{time in hours} &= \frac{10\ 577.777}{3600} \\
&= 2.938
\end{aligned}
$$

It would take approximately 3 hours to heat the water.

If the cost of 1 kWh is 7p, then the cost of heating the water would be:

2.938 × 3 × 7 = 61.7p (approx.)

Example 6.22. *Calculate the percentage efficiency of an electric water heater having a capacity of 250 litres when its heating element is 6 kW and the time* *taken to raise the temperature of water from $5^{\circ}C$ to $60^{\circ}C$ is 3 hours.*

$$\text{power} = \frac{\text{s.h.c.} \times \text{kg} \times \text{temperature rise (K)} \times 100}{\text{heating up time in seconds} \times \text{efficiency}} \text{ kW}$$

by transposition,

$$\text{efficiency} = \frac{\text{s.h.c.} \times \text{kg} \times \text{temperature rise (K)} \times 100}{\text{heating up time in seconds} \times \text{kW}}$$

For practical purposes, the mass of 1 litre of water may be taken as 1 kg.

$$\therefore \quad \text{efficiency} = \frac{4.2 \times 250 \times (60 - 5) \times 100}{3 \times 3600 \times 6}$$

$$= 89.12 \text{ per cent}$$

Example 6.23. *Calculate the power in kW required for an immersion heater element when it is required to raise the temperature of 73 litres of water from $10^{\circ}C$ to $60^{\circ}C$ in 2.5 hours and the heat losses are assumed to be 20 per cent.*

$$\text{power} = \frac{\text{s.h.c.} \times \text{kg} \times \text{temperature rise (K)} \times 100}{\text{heating up time in seconds} \times \text{efficiency}} \text{ kW}$$

$$= \frac{4.2 \times 73 \times 50 \times 100}{2.5 \times 3600 \times 80} \text{ kW}$$

$$\doteq 2.13 \text{ kW}$$

A 2 kW heater would be satisfactory.

Pumping

The power required for pumping is found as follows:

$$\text{power} = \frac{\text{work done}}{\text{time}}$$

$$= \frac{\text{force} \times \text{distance}}{\text{time}}$$

$$= \frac{\text{newtons} \times \text{distance}}{\text{time}}$$

$$\text{force in newtons} = \text{mass} \times 9.81$$

$$\text{power in watts} = \frac{\text{mass} \times 9.81 \times \text{distance}}{\text{time}}$$

Example 6.24. *A centrifugal pumping set is required to raise 5 kg/s of water vertically to a height of 20 m. Allowing 30 per cent of the height for pipe friction, calculate the power in kilowatts required if the efficiency of the equipment is 70 per cent.*

$$\text{effective height} = 20 + (20 \times 0.3) = 26 \text{ m}$$

$$\text{power in watts} = \frac{\text{newtons} \times \text{metres}}{\text{seconds}}$$

$$= \frac{kg \times 9.81 \times metres}{seconds}$$

$$= \frac{5 \times 9.81 \times 26}{1}$$

$$= 1275.3$$

Allowing 70 per cent efficiency,

$$actual\ power = 1275.3 \times \frac{100}{70}$$

$$= 1821.857\ watts$$

Pump laws

When the impeller diameter is constant the following laws apply:

1. The discharge varies directly with the angular velocity of the impeller.
2. The pressure developed varies as the square of the angular velocity of the impeller.
3. The power absorbed varies as the cube of the angular velocity of the impeller.

The laws may be written:

1. $\dfrac{q_2}{q_1} = \dfrac{N_2}{N_1}$

2. $\dfrac{P_2}{P_1} = \dfrac{(N_2)^2}{(N_1)^2}$

3. $\dfrac{W_2}{W_1} = \dfrac{(N_2)^3}{(N_1)^3}$

where

q_1 and q_2	= discharge in litres per second
P_1 and P_2	= pressure in kPa
N_1 and N_2	= angular velocity, revolutions per minute
W_1 and W_2	= power absorbed in watts

Example 6.25. *A centrifugal pumping set absorbs 2 kW and discharges 4 litre/s when the angular velocity of the impeller is 800 revolutions per minute. If the angular velocity of the impeller is increased to 1000 revolutions per minute, calculate: (a) new discharge, (b) new power absorbed.*

(a) $\dfrac{q_2}{q_1} = \dfrac{N_2}{N_1}$

$q_2 = \dfrac{q_1 \times N_2}{N_1} = \dfrac{4 \times 1000}{800}$

$= 5\ litre/s$

(b) $\dfrac{W_2}{W_1} = \dfrac{(N_2)^3}{(N_1)^3}$

$W_2 = \dfrac{W_1 \times (N_2)^3}{(N_1)^3} = \dfrac{2000 \times 1000^3}{800^3}$

$= 3906.25\ watts$

Example 6.26 (see Fig. 6.8).

Fig.6.8 Pumping in high-rise building

A cold-water storage cistern is 60 m vertically above a centrifugal pump which is to be used for raising 6000 litres of water from a low-level break cistern every 3 hours. Allowing 30 per cent of the height for pipe friction, calculate the rating of the electric motor if the efficiency of the pumping equipment is 65 per cent.

effective height = $60 + (60 \times 0.3) = 78$ m

$$\text{discharge} = \frac{6000}{3 \times 60 \times 60} = 3.333 \text{ litre/s}$$

$$\text{actual power} = \frac{\text{newtons} \times \text{metres}}{\text{seconds}} \times \frac{100}{\text{efficiency}}$$

$$= \frac{3.333 \times 9.81 \times 78}{1} \times \frac{100}{65}$$

$$= 3923.6075 \text{ watts}$$

required rating of electric motor = 4 kW (approx.)

Pumping definitions

Capacity: the flow rate discharged by a pump, usually expressed in m^3 /h or l/s.

Static head: the vertical height of the liquid being pumped.

Velocity pressure: the pressure needed to set the liquid in motion.

Friction head: the head necessary to overcome resistance to flow of the liquid. Friction head must not exceed the total delivery head.

Negative suction lift: this exists when the pump is above the liquid to be pumped and is the vertical distance from the centre line of the pump down to the free surface of the liquid.

Positive suction head: this exists when the pump is below the liquid to be pumped and is the vertical distance from the centre-line of the pump up to the free surface of the liquid.

Static delivery head: the vertical distance between the centre-line of the pump and the free surface of the discharged liquid.

Total negative suction lift: the static suction lift plus the friction head and velocity head in the suction pipe system.

Total positive suction head: the static suction head minus the friction head and velocity head in the inlet pipe system.

Total delivery head: the static delivery head plus the friction head and velocity head in the delivery pipe system.

Fig. 6.9 (a) with negative suction lift; (b) with flooded or positive suction head

101

(b)

Static delivery head

Flooded or positive suction head

Motor   Pump

Conductor and cable rating

The amount of current which a conductor or cable can carry is limited by the heating effect caused by the resistance to flow of electricity. The maximum permissible current under normal conditions must not be so high that dangerous temperatures may be reached, which could lead to fires. Even with cables inside metal conduits or ducts, or where mineral insulated copper or aluminium sheathed cables are used, although the cables are completely fireproof in themselves, the transmission of heat to other materials in proximity may still lead to fires.

When choosing a cable for a particular duty, it is necessary to take into account not only the maximum current the cable will have to carry, but also the voltage drop that will occur when the current flows.

The 15th edition of the Institute of Electrical Engineers (IEE) Regulations state that the maximum permissible drop in voltage in a conductor shall not exceed 2.5 per cent of the nominal voltage when the conductor is carrying its full load.

The temperature reached by a cable is also affected by the following operating conditions:

1. Whether the cable is surrounded by room air, or is enclosed in a conduit or duct.
2. The proximity to other cables which may cause heat to build up, due to induced currents.
3. The temperature of the ambient or surrounding air.

The 15th edition of the IEE regulations provide tables of cable sizes for various operating conditions (see Table 6.1).

Example 6.28. *A 25 m run of two-core non-armoured PVC insulated copper cable, enclosed in conduit, is required to supply a 10 kW load. If the nominal voltage of the supply is 240, find the size of cable when coarse excess current protection is to be provided.*

$$\text{current} = \frac{W}{V} = \frac{10\ 000}{240} = 41.667\ \text{A}$$

Table 6.1 (Part Table 9DI of IEE Regulations)   Single-circuit rating and associated voltage drops for single core PVC insulated cables, non-armoured with copper conductors. The ratings tabulated apply where the cable is provided with coarse excess-current protection.

Conductor		Bunched and enclosed in conduit or trunking	
Nominal cross-sectional area (mm²)	Number and diameter of wires (mm)	Two cables single phase a.c. or d.c.	
		Current rating (A)	Voltage drop per ampere per metre run
1.0	1/1.13	14	42
1.5	1/1.38	17	28
2.5	1/1.78	24	17
4	7/0.85	32	11
6	7/1.04	41	7.1
10	7/1.35	55	4.2

From Table 6.1, a conductor having a cross sectional area of 10 mm² will allow a current of 55 A to flow with a voltage drop of 4.2 mV or 0.0042 V per ampere per metre run.

$\therefore$ total voltage drop = 0.0042 × 41.667 × 25 in cable
= 4.375 V

check on percentage of nominal voltage:

$$\frac{4.375}{240} \times \frac{100}{1} = 1.823 \text{ per cent}$$

This is below 2.5 per cent of the nominal voltage and the cable is therefore suitable.

Gas installations

Flow of gas in pipes

The flow of gas in pipes will depend upon the following factors:

1. the internal diameter of the pipe;
2. the pressure of gas;
3. the type of pipe material used;
4. the relative density of the gas;
5. the length of pipe and resistances of fittings.

Normally the pressure of North Sea gas to the premises is up to 5 kPa, which will be governed down to give a pressure of between 2 to 2.25 kPa in the installation pipework.

Resistances of fittings

In order to find the effective length of pipe, the resistances of fittings must be added to the net length. Table 6.2 gives the additions for fittings.

Table 6.2 Additions to net pipe run (m)

Nominal bore	Elbows	Tees	90° bends
15–25	0.6	0.6	0.3
32–40	1.0	1.0	0.3
50	1.6	1.6	0.3

Rate of flow: Manufacturers of gas appliances usually give the rate of flow in m^3/h.

Table 6.3 gives typical gas consumption at appliances.

Table 6.3 Gas consumption of appliances

Appliance	Consumption (m^3/h)
Refrigerator	0.10
Wash boiler	1.13
Sink water heater	2.30
Instantaneous multi-point water heater	5.70
Cooker	3.70
Warm-air heater	2.30

Table 6.4 gives the discharge rates through steel pipes; it may be used in conjunction with Tables 6.1 and 6.2.

Table 6.4 Discharge rates through steel pipes

Nominal bore (mm)	Length of pipe in metres									
	3	6	9	12	15	18	21	24	27	30
6	0.1									
8	0.1	0.3								
10	1.6	1.3	0.85							
15	3.0	2.1	1.17							
20	6.5	4.5	3.7	3.0	2.8	2.5				
25	13.3	9.4	7.6	6.8	6.0	5.4	5.0	4.8	4.5	
32	26.3	18.4	15.3	13.3	11.9	10.7	10.0	9.3	8.7	8.2
40	26.8	26.0	20.7	18.4	16.4	15.0	13.8	13.0	12.0	11.6
50	73.6	53.8	42.5	36.8	34.0	31.0	28.0	26.3	25.0	23.8

Discharge (m^3/h)

Example 6.29. *Find the internal diameter of a gas pipe to supply one sink water heater, two instantaneous multi-point water heaters, two cookers and* *one refrigerator. The net length of pipe required is 16 m and there will be four 90° bends and five tees in the pipe run.*

Gas consumption (Table 6.3)

sink water heater		=	2.30 m³/h
instantaneous multi-point			
water heaters	= 2 × 5.70 =		11.40 m³/h
cookers	= 2 × 3.70 =		7.40 m³/h
refrigerator		=	0.10 m³/h
		total	21.20 m³/h

Resistance of pipe fittings assuming a 50 mm diameter pipe (see Table 6.2)

90° bends	= 4 × 0.6 =	2.4 m
tees	= 5 × 1.6 =	8.0 m
	total	10.4 m

effective length of pipe = 16 + 10.4 = 26.4 m

Table 6.4 shows that a 50 mm diameter pipe will discharge 25 m³/h when the length of pipe is 27 m and is therefore adequate.

Pole's formula

The flow of gas in pipes may be found from Pole's formula, which is expressed as follows:

$$q = 0.001\,978\,d^2 \sqrt{\frac{H\,d}{S\,L}}$$

where
q = litres per second
d = diameter of pipe in mm
H = pressure drop in mb
S = relative density of gas (0.58–0.59)
L = length of pipe in m

Calculation of gas consumption

The gas consumed may be found from the following formula:

$$\text{consumption} = \frac{\text{power} \times \text{seconds}}{\text{calorific value}}\ \text{m}^3/\text{s}$$

The calorific value or heating value of North Sea gas is approximately 37 MJ/m³, and that of town gas is approximately 19 MJ/m³.

If the gas appliance is rated in kW this will have to be converted into kJ as follows:

$$\text{power} = \frac{\text{kJ}}{\text{seconds}}\ \text{kW}$$

∴   kJ = kW × seconds

Example 6.30. *Calculate the gas consumed in m^3/h by a 25 kW heating boiler when North Sea gas is used.*

consumption $= \dfrac{25 \times 3600}{37\,000} = 2.432 \text{ m}^3/h$

Note: 1 kW = 3.6 MJ/h
1 therm = 105.5 MJ

Example 6.31. *Calculate the cost of operating a 30 kW gas-fired boiler for 84 hours when 1 therm costs 40p.*

consumption per hour $= $ kW $\times$ 3.6 $\quad = 108$ MJ
$\therefore$ MJ for 84 hours $\quad\;\; = 108 \times 84 \quad = 9072$ MJ

$\quad$ cost for 84 hours $\quad = \dfrac{9072}{105.5} \times \dfrac{40}{1} = 3439.62$p

$\qquad\qquad$ cost $\quad = £34.40$ (approx.)

Alternatively, the consumption may be found as follows:

kW $= \dfrac{\text{kJ}}{\text{s}}$

kJ $=$ kW $\times$ s
$\quad = 30 \times 84 \times 60 \times 60$
$\quad = 9\,072\,000$

$\therefore$ MJ $= \dfrac{9\,072\,000}{1000} = 9072$

Example 6.32. *Calculate the amount of North Sea gas consumed in m^3 and the cost of heating 136 litres of water from $10\,^\circ C$ to $60\,^\circ C$ by use of a gas circulator having a power input of 3 kW when the overall efficiency is 70 per cent, the cost of 1 therm is 30p, the calorific value of gas is 37 MJ/m^3, and the specific heat capacity (s.h.c.) of water is 4.2 kJ/kg K.*

$\qquad\qquad$ power $= \dfrac{\text{s.h.c.} \times \text{kg} \times \text{temperature rise K} \times 100}{\text{heating time in seconds} \times \text{efficiency}}$ kW

$\quad$ time in seconds $= \dfrac{\text{s.h.c.} \times \text{kg} \times \text{K rise} \times 100}{\text{kW} \times \text{efficiency}}$

$\qquad\qquad\qquad = \dfrac{4.2 \times 136 \times 50 \times 100}{3 \times 70}$

$\qquad\qquad\qquad = 13\,600$ seconds

$\quad$ time in hours $= \dfrac{13\,600}{3600}$

$\qquad\qquad\qquad = 3.778$ h

$\qquad$ MJ per h $=$ kW $\times$ 3.6

$\qquad\qquad\qquad = 3 \times 3.6 = 10.8$ MJ

MJ for 3.778 h $= 10.8 \times 3.778$

$\qquad\qquad\qquad = 40.802$ MJ

$$\text{therms used} = \frac{40.802}{105.5}$$

$$= 0.3867$$

$$\text{cost at 40p per therm} = 0.3867 \times 40$$

$$= 15.5\text{p}$$

$$\text{gas consumed} = \frac{\text{kW} \times \text{seconds used}}{\text{calorific value kJ/m}^3}$$

$$= \frac{3 \times 13\ 600}{37\ 000}$$

$$= 1.103 \text{ m}^3$$

Example 6.33. *Calculate the operating cost of a heating system for one season if the building has a steady design heat loss of 300 kW. Use the following factors:*

(a) *Heating season* = *30 weeks*
(b) *Hours of heating per 24 hours* = *10*
(c) *Boiler efficiency* = *70 per cent*
(d) *Load factor* = *50 per cent*
(e) *Calorific value of gas* = *37 MJ/m³*
(f) *Cost of 1 therm* = *40p*

$$\begin{aligned}\text{power required allowing} \\ \text{50 per cent load factor} &= \text{50 per cent of 300 kW}\end{aligned}$$

$$= 150 \text{ kW}$$

$$\text{time in seconds} = 30 \times 7 \times 10 \times 3600$$

$$= 7\ 560\ 000$$

$$\text{kJ} = \text{kW} \times \text{seconds}$$

$$= 150 \times 7\ 560\ 000$$

$$= 1\ 134\ 000\ 000$$

$$\text{MJ} = 1\ 134\ 000$$

$$\text{1 therm} = 105.5 \text{ MJ}$$

$$\therefore \text{therms} = \frac{1\ 134\ 000}{105.5}$$

$$= 10\ 748.815$$

$$\text{cost (£)} = \frac{10\ 748.815 \times 40 \times 100}{100 \times 70}$$

$$= £6142.18$$

$$\text{Alternatively, MJ/h} = \text{kW} \times 3.6$$

$$= 150 \times 3.6$$

$$= 540$$

$$\text{MJ for 30 weeks} = 540 \times 10 \times 7 \times 30$$

$$= 1\ 134\ 000$$

$$\text{cost } (£) = \frac{1\ 134\ 000 \times 40 \times 100}{105.5 \times 100 \times 70}$$

$$= £6142.18$$

Note: The steady design heat loss will be based on an outside air temperature of $-1\,°C$ or $-2\,°C$, therefore a load factor is introduced for a heating season in order to take into account the varying outside air temperatures. For a shorter period of, say, one week, the outside air temperature may have a mean of about $-1\,°C$ or $-2\,°C$ and therefore the load factor will be 100 per cent or unity.

Gas pressure

If the gas pressure is given in kilopascals it is sometimes necessary to convert this pressure into millibars or the equivalent reading in millimetres water gauge.

Example 6.34. *The pressure of gas required for an appliance burner is given as 2 kPa. Convert this pressure to millibars pressure and the equivalent reading in millimetres water gauge.*

$$1\ \text{mbar} = 100\ \text{N/m}^2$$
$$2\ \text{kPa} = 200\ \text{N/m}^2 \text{ or } 2000\ \text{Pa}$$

$$\text{mbars} = \frac{2000}{100}$$

$$= 20\ \text{mbars}$$

$$\text{pressure in pascals} = \text{head in metres} \times 1000 \times 9.81$$

$$\text{head in metres} = \frac{\text{pressure in pascals}}{1000 \times 9.81}$$

$$= \frac{2000}{1000 \times 9.81}$$

$$= 0.203\ 873\ 5$$

$$\text{head of water in millimetres} = 203.8735$$

For practical purposes, a manometer reading of 204 mm water gauge would be satisfactory (see Fig. 6.10).

Gas cock open

Rubber tube

Installation pipe

Open end

Head of water in mm   204 mm

Gas pressure forces water up the tube

Glass U tube or manometer

Fig.6.10 Gas pressure

Questions

1. With a supply voltage of 240 V, calculate: (a) the current taken by a 2 kW electric immersion heater; (b) an 80 W lamp.

Answers: (a) 8.333 A; (b) 0.333 A

2. Calculate the maximum power load which may be connected to a 30 A ring circuit when the voltage is 240.

Answer: 7200 W or 7.2 kW

3. (a) The current in a circuit is 60 A and the resistance 0.2 Ω. Calculate the voltage.
   (b) The voltage applied to a circuit is 110 V and the current flowing is 13 A. Calculate the resistance.

109

(c) If a voltage of 240 V is connected across a resistor of 100 Ω, what current will flow?

Answers: (a) 12 V; (b) 8.461 Ω; (c) 2.4 A

4. Calculate the current flowing in and the resistance of the element of a 5 kW immersion heater (when in use) when connected to a 240 V supply.

Answers: 20.833 A; 11.520 Ω

5. State Ohm's law. Two resistors of 5 Ω and 8 Ω respectively are connected in series, and a potential difference of 12 V is applied to the circuit. Draw the circuit and calculate: (a) the total current flowing; (b) the voltage across each resistor.

Answers: (a) 0.923 A; (b) 4.615 V and 7.384 V

6. The electric lighting of a classroom consists of 12 fluorescent lamps, each lamp requiring 100 W at 0.85 power factor. If the supply is 240 V single phase, calculate: (a) total power in kW; (b) cost of running the lamps for 8 hours when 1 kWh cost 5p; (c) total kVA; (d) current supplied to the circuit and fuse rating in amperes.

Answers: (a) 1.2 kW; (b) 48p; (c) 1.412; (d) 58.823 A, fuse rating 59 A

7. A 3 kW motor on a 240 V single-phase supply is found to take 18 A and 3.8 kW on full load. Calculate: (a) the power factor; (b) the percentage efficiency of the motor.

Answers: (a) 0.88; (b) 78.947 per cent efficiency

8. Calculate the time taken in hours and the cost of heating 360 kg of water from 5 °C to 60 °C by means of a 8 kW immersion heater when the heat losses are 20 per cent and the cost of 1 kWh is 7p.

Answers: 3.609 hours; 202.104p.

9. Two 240 V electric lamps, 100 W and 150 W respectively, are connected in parallel to a 240 V supply. Calculate: (a) the current flowing through each lamp; (b) the resistance of each lamp when in use; (c) the total current from the supply.

Answers: (a) 0.4166 A and 0.625 A; (b) 576 Ω and 384 Ω; (c) 1.0416 A

10. A centrifugal pump is required to raise water through a vertical height of 20 m at the rate of 15 kg/s. Allowing for a friction loss on the pipework of 2 m head, calculate the power rating of the motor in kW if the overall efficiency of the pumping unit is 70 per cent.

Answer: 4624.714 W (= 4.6 kW approx.)

11. Define: (a) kilovolt-amperes; (b) power factor; (c) root mean square value; (d) kilowatts; (e) kilowatt hours.

12. A load of 6 kW is to be supplied at 240 volts a.c. through a two-core cable having a resistance per conductor of 0.0025 ohms per metre and the cable length is 10 metres. Find: (a) the size of fuse for the circuit; (b) the voltage drop over the total length of cable.

Answers: (a) 25 A; (b) 0.0625 V

13. Calculate the cost of operating a 25 kW gas-fired boiler for 24 hours when 1 therm costs 40p.

Answer: £8.09 (approx.)

14. Calculate the volume of gas consumed in m^3 and the cost of heating 227 litres of water from 5 °C to 60 °C by use of a gas storage-type water heater having a power input of 5 kW when the overall efficiency of the heater is 75 per cent.
    Cost of 1 therm = 40p; calorific value of gas = 37 000 kJ/m^3

Answers: 1.34 m^3; 26.52p

15. Calculate the operating cost of a heating system for one week if the building has a steady design heat loss of 50 kW. Use the following factors:
    (a) Hours of heating per day   = 12 hours;
    (b) Boiler efficiency   = 80 per cent;
    (c) Load factor   = 100 per cent;
    (d) Calorific value of gas   = 37 MJ/m^3;
    (e) Cost per therm   = 40p

Answer: £71.659

16. If a manometer connected to a gas service pipe reads 210 mm, calculate the pressure in millibars.

Answer: 20.6 mbars

Chapter 7

Heating

Heat-loss calculations

Before showing the methods of calculating heat losses from buildings it is necessary to define the terms used.

Definition of terms

1. **Thermal conductivity (k) (see Fig. 7.1)** The thermal transmission in unit time through unit area of a slab, or a uniform homogeneous material of unit thickness, when unit difference of temperature is established between its surfaces. The unit is W/m K.

    The definition is based on the experimental formula for rate of flow of heat.

$$\frac{Q}{t} = \frac{kA\,(\theta_1 - \theta_2)}{d}$$

where Q = quantity of heat, measured in joules, passing through the material in a fixed time

t = time in seconds

k = thermal conductivity

θ_1 and θ_2 = temperature of each face of the material (°C)

A = the cross-sectional area of the material through which the heat is passing (m^2)

d = thickness of the material (m)

θ(2°C)

Warm

d = 1 metre

Uniform homogeneous
building material

Area 1 metre²

θ(1°C)

Heat flow joules per
second

Cold

Fig.7.1 Thermal conductivity (k value)

In SI units, thermal conductivity is expressed as heat flow in watts per square metre of surface area, for a temperature difference of 1 °C per metre thickness.

If the units are substituted in the above equation, the units of k will be:

$$k = \frac{J\ m}{m^2\ s\ K}$$

since $W = \dfrac{J}{s}$, this becomes $\dfrac{W\ m}{m^2\ K}$

but thickness over area m/m² cancels to 1/m, so that k becomes W/m K.

Table 7.1 gives the k values of common types of building materials.

2. **Thermal resistivity** (r)   This is the reciprocal of the thermal conductivity and is used for calculating the total resistance or the conductance of a material. The unit is m K/W (note m²/m cancels to m).

When the thickness of the material is known, its thermal resistance (R) can be calculated: this is the product of thermal resistivity ($1/k$) and thickness and is expressed as m² K/W. The thickness of the material must be in metres.

3. **Thermal conductance** (c)   This is the thermal transmission in unit time, through a unit area of a uniform structural component of thickness L per unit of temperature difference between the hot and cold surfaces. The unit is W/m² K.

4. **Thermal resistance** (R)   This is the reciprocal of thermal conductance:

$$R = \frac{L}{c}$$

where L is expressed in metres.

113

Table 7.1 Thermal conductivities (k) of common building materials

Materials	k value (W/m K)
Asphalt	1.20
Common brick	1.20
Dense brick	1.47
Aerated concrete	0.14
Structural concrete	1.40
Clinker block	0.05
Cork slab	0.40
Glass wool	0.034
Glass	1.02
Mineral wool	0.037
Gypsum plasterboard	0.15–0.58
Gypsum plaster	0.40
Polyurethane foam	0.020–0.025
Rendering (cement and sand)	0.53
Sandstone	1.30
Limestone	1.50
Clay tiles	0.83
Softwood	0.138
Wood-wool slabs	0.09
Chip board	0.108
Air	0.029
Soil	1.00–1.15
Vermiculite	0.065

For two or more materials in the same structure, the resistance of the individual materials may be added to obtain the resistance in the structure.

total resistance = $R_1 + R_2 + R_3 + \dots$

5. Thermal transmission (U) This is the thermal transmission in unit time through unit area of a given structure. The unit is W/m^2 K.

The thermal transmission through the structure is obtained by combining the thermal resistances of its components and the adjacent air layers. The thermal transmission therefore is found by adding the thermal resistances and taking the reciprocal.

$$U = \frac{1}{R_{si} + R_{so} + R_1 + R_2 + R_3 + R_a}$$

where U = thermal transmission (W/m^2 K)
R_{si} = inside surface resistance (m^2 K/W)
R_{so} = outside surface resistance (m^2 K/W)
R_1, R_2, R_3 = thermal resistance of structural component (m^2 K/W)
R_a = resistance of air space (m^2 K/W)

In computation of U values the thermal resistance L/k is used.

where k = thermal conductivity (W/m K)
L = thickness in metres of a uniform homogeneous material

$$\therefore U = \frac{1}{R_{si} + R_{so} + L_1/k_1 + L_2/k_2 + L_3/k_3 + R_a}$$

Surface-air effects (see Fig. 7.2)

A layer of stationary air is formed on the inside and outside surfaces of the structure. These layers of stationary air act as insulators and must therefore be included in the calculation to find the total resistance of the structure.

The symbols used to denote these surface resistances in the above formula are R_{si} and R_{so}.

Typical values for these resistances (m^2 K/W) are:

horizontal R_{si} = 0.123
horizontal R_{so} = 0.053
upwards R_{si} = 0.110
downwards R_{si} = 0.150
upwards R_{so} = 0.045

Fig.7.2 Surface-air effects

Resistance of air space

Heat may be transmitted in a cavity by convection, conduction and radiation. If a cavity is less than 20 mm wide it is usual to consider the air as stationary and to deal with it as an extra layer of insulating material. If the cavity is greater than 20 mm wide it is usual to take the resistance as being about 0.18 m^2 K/W.

Calculation of *U* values

In order to calculate the thermal transmission through a structure, it is preferable to make a sketch of the structure showing the various components and their thicknesses.

Example 7.1. *Figure 7.3 shows a section through a solid brick wall. Using the following data, calculate the thermal transmission (U) for the wall.*

Fig.7.3 Solid brick wall (U value = 2.3 W/m² K)

Thermal conductivities (W/m K)

brick	1.20
cement and sand render	0.53
plaster	0.40

Surface resistances (m² K/W)

R_{si} internal surface layer	0.123
R_{so} external surface layer	0.053

$$U = \frac{1}{R_{si} + R_{so} + L_1/k_1 + L_2/k_2 + L_3/k_3}$$

$$= \frac{1}{0.123 + 0.053 + 0.019/0.40 + 0.225/1.20 + 0.015/0.53}$$

$$= \frac{1}{0.123 + 0.053 + 0.0475 + 0.1875 + 0.0283}$$

$$= \frac{1}{0.4393}$$

$$= 2.276 \text{ W/m}^2 \text{ K}$$

$$= 2.3 \text{ (approx.)}$$

Example 7.2. *Figure 7.4 shows a section through a cavity wall. Using the following data, calculate the thermal transmission (U) for the wall.*

Thermal conductivities (W/m K)

brick	1.20
aerated concrete	0.14
plaster	0.40

116

Fig.7.4 Cavity wall (U value = 0.934 W/m² K)

Surface resistances (m² °C/W)

R_{si} internal surface layer	0.123
R_{so} external surface layer	0.053
R_a air space	0.18

$$U = \frac{1}{R_{si} + R_{so} + L_1/k_1 + L_2/k_2 + L_3/k_3 + R_a}$$

$$= \frac{1}{0.123 + 0.053 + 0.08/0.14 + 0.115/1.20 + 0.019/0.40 + 0.18}$$

$$= \frac{1}{0.123 + 0.053 + 0.571 + 0.096 + 0.0475 + 0.18}$$

$$= \frac{1}{1.0705}$$

$$= 0.934 \text{ W/m}^2 \text{ K}$$

Example 7.3. *If the cavity of the wall given in example 7.2 is filled with polyurethane foam, having a thermal conductivity value of 0.022 W/m K, calculate the U value for the wall and the percentage saving in heat loss. Instead of using 0.18 m² K/W for the resistance of the air space, the resistance of 50 mm thickness of polyurethane foam will be used.*

$$\text{resistance of cavity fill} = \frac{L_4}{k_4} = \frac{0.05}{0.022} = 2.273$$

From the answer given in example 7.2,

117

$$U = \frac{1}{0.123 + 0.053 + 0.571 + 0.096 + 0.0475 + 2.273}$$

$$= \frac{1}{3.1634}$$

$$= 0.316 \text{ W/m}^2 \text{ K}$$

The percentage saving in heat loss through the wall by filling the cavity with polyurethane foam would be:

difference $= 0.934 - 0.316 = 0.618 \text{ W/m}^2 \text{ K}$

saving $= \dfrac{0.618}{0.934} \times \dfrac{100}{1} = 66.167$ per cent

---

With the high cost of fuel, the saving by filling the cavity should ultimately pay for the capital cost of the additional insulation.

Building Regulations 1985 (Conservation of Fuel and Power)

Requirement for dwellings.
1. The calculated rate of heat loss (W K) through any windows and roof lights shall be no greater than it would be if:
   (a) the aggregate of the areas of windows and roof lights were 12 per cent of the area of the walls bounding them; and
   (b) the windows and roof lights had a U value of 5.7.

2. The calculated rate of heat loss through the solid parts of the exposed elements shall be not greater than it would be if:
   (a) the exposed walls and exposed floors had a U value of 0.6; and
   (b) the roof had a U value of 0.35.

3. The extent that the calculated rate of heat loss through the solid parts of exposed elements is less than the maximum permitted under sub-paragraph (2) the calculated rate of heat loss through windows and roof lights may be greater than the maximum permitted under paragraph (1).

Table 7.2 gives the maximum U value required under the regulations which apply to any building, or part of a building, which is intended to be used as a dwelling. Other buildings are included.

The regulations do not apply to any external wall, floor, or roof of any part of a dwelling which consists of a shed or store entered from outside or of a garage, boathouse, conservatory or porch.

Permanent vent means an opening or duct which communicates with the external air and is designed to allow a passage of air at all times. U value means thermal transmission coefficient, that is to say, the rate of heat transfer in watts through 1 m^2 of a structure when the combined radiant and air temperatures at each side of the structure differ by 1 $^\circ$C, and it is expressed in W/m^2 $^\circ$ K.

Table 7.2 Maximum U value of walls, floors and roofs

Type of building	Exposed walls	Roofs	Exposed floors
Dwellings	0.6	0.35	0.6
Other residential buildings including hotels and institutions	0.6	0.6	0.6
Places of assembly offices and shops	0.6	0.6	0.6
Industrial and storage	0.7	0.7	0.7

Note: Display windows in shops are not included.

Minimum thickness of components

In order to comply with the Building Regulations it is sometimes necessary to calculate the minimum thickness of a component forming part of the structure.

Example 7.4. *An external cavity wall is to be constructed so that its U value does not exceed 1 W/m² K. Calculate the minimum thickness of aerated concrete inner leaf of the wall when the brick outer leaf is 115 mm thick; cavity 50 mm wide and internal plastering 19 mm thick.*

Use the following factors:

Thermal conductivities (W/m K)
brick	1.20
aerated concrete	0.14
plaster	0.40

Surface resistances (m² K/W)
R_{si} internal surface layer	0.123
R_{so} external surface layer	0.053
R_a air space	0.18

$$U = \frac{1}{R_{si} + R_{so} + L_1/k_1 + L_2/k_2 + L_3/k_3 + R_a}$$

$$\frac{1}{U} = R_{si} + R_{so} + L_1/k_1 + L_2/k_2 + L_3/k_3 + R_a$$

Taking the thickness of the inner leaf as L_2 and transposing:

$$1 - R_{si} - R_{so} - \frac{L_1}{k_1} - \frac{L_3}{k_3} - R_a = \frac{L_2}{k_2}$$

$$L_2 = (1 - R_{si} - R_{so} - \frac{L_1}{k_1} - \frac{L_3}{k_3} - R_a) \times k_2$$

$$= (1 - 0.123 - 0.053 - \frac{0.115}{1.2} - \frac{0.019}{0.4} - 0.18) \times 0.14$$

$$= (1 - 0.123 - 0.053 - 0.095\,83 - 0.0475 - 0.18) \times 0.14$$

$$= 0.070\,094 \text{ m}$$

$$= 70 \text{ mm (approx.)}$$

Check:

$$U = \frac{1}{0.123 + 0.053 + 0.0958 + 0.07/0.14 + 0.0475 + 0.18}$$

$$= \frac{1}{0.123 + 0.053 + 0.0958 + 0.5 + 0.0475 + 0.18}$$

$$= \frac{1}{0.9993}$$

$$= 1.000 \text{ W/m}^2 \text{ K}$$

The 70 mm thickness of aerated concrete inner leaf would satisfy the minimum required but an additional saving in fuel would be achieved by a 100 mm thickness.

Heat losses through pitched roofs (see Fig. 7.5)

In order to calculate the U value for pitched roofs with a horizontal ceiling below, the following equation may be used:

$$U = \frac{U_r \times U_c}{U_r + U_c \cos a}$$

where U = the combined thermal transmission for the ceiling and roof (W/m^2 K)

U_r = the thermal transmission for the roof (W/m^2 K)

U_c = the thermal transmission for the ceiling

a = the angle of the pitched roof

Fig.7.5 Thermal transmission (U value) for pitched roofs

Example 7.5. *A pitched roof inclined at 40° has a horizontal ceiling below. If* U_r *and* U_c *are found to be 2.50 W/m² K and 0.9 W/m² K respectively, calculate the combined U value for the ceiling and roof.*

$$U = \frac{2.5 \times 0.9}{2.5 + (0.9 \times 0.776)}$$

$$= 0.70 \text{ W/m}^2 \text{ K (approx.)}$$

Environmental temperature (t_e)

This is a balanced mean, between the mean radiant temperature and the air temperature. It may be evaluated approximately from the following formula:

$$t_{ei} = \frac{2}{3} t_{ri} + \frac{1}{3} t_{ai}$$

where t_{ri} = mean radiant temperature of all room surfaces (°C)

t_{ai} = inside air temperature (°C)

The approximate value of the mean radiant temperature may be found by totalling the products of the various areas and the surrounding surfaces and dividing this total by the sum of the areas; i.e.,

$$t_{ri} = \frac{a_1 t_1 + a_2 t_2 + a_3 t_3 + \ldots}{a_1 + a_2 + a_3 + \ldots}$$

where

t_1, t_2, t_3 = surface temperature (°C)

a_1, a_2, a_3 = surface areas (m²)

Table 7.3 provides a summary of recommended environmental temperatures.

Table 7.3 Summary of recommended internal (winter) environmental temperatures

	t_{ei} (°C)		t_{ei} (°C)
Art galleries	20	Laboratories	20
Assembly halls	18	Law courts	20
Bars	18	Libraries	20
Canteens	20	*Offices*	
Churches	18	General	20
Factories		Private	20
Sedentary work	19	Police stations	18
Light work	16	Restaurants	18
Heavy work	13	*Hotels*	
Flats and houses		Bedrooms (standard)	22
Living rooms	21	Bedrooms (luxury)	24
Bedrooms	18	Public rooms	21
Bathrooms	22	*Schools and colleges*	
Entrance hall	16	Classrooms	18
Hospitals		Lecture rooms	18
Corridors	16	*Shops*	
Offices	20	Small	18
Operating theatre	18–21	Large	18
Wards	18	Swimming baths	
Sports pavilions	21	Changing rooms	22
Warehouses	16	Bath hall	26

Example 7.6 *Using the following data, calculate the environmental temperature of a room in which the air temperature is 22 °C.*

Fabric	Area (m^2)	Surface temperature (°C)
Floor	200	18
Walls	180	20
Windows	60	10
Ceiling	200	16

$$t_{ri} = \frac{(200 \times 18) + (180 \times 20) + (60 \times 10) + (200 \times 16)}{200 + 180 + 60 + 20}$$

$$= \frac{3600 + 3600 + 600 + 3200}{460}$$

$$= \frac{11\,000}{460}$$

$$= 23.9 \,°C$$

The environmental temperature may be found as follows:

$$t_{ei} = \frac{2}{3} t_{ri} + \frac{1}{3} t_{ai}$$

$$= \left(\frac{2}{3}\right) \times 23.9 + \left(\frac{1}{3}\right) \times 22$$

$$= 15.933 + 7.333$$

$$= 23.266\,^{\circ}C$$

$$= 23.3\,^{\circ}C \text{ (approx.)}$$

The concept of the environmental temperature for heat-loss calculations provides a more accurate assessment of steady-state heat loss through a structure, than does the conventional method that uses internal air temperature, only, as a basis.

The internal environmental temperature is also better than the internal air temperature as an index of the internal comfort of the internal environment. This places the designer in a more favourable position for assessing the internal thermal comfort than is possible by employing air temperature.

The conventional method of using the internal air temperature may be used when the difference between the mean radiant temperature and the air temperature is quite small. This occurs when rooms have little exposure to the outside and the standard of thermal insulation is very high.

The thermal transmission (U) values of common forms of construction may be obtained from various organisations. Table 7.4 provides a list of typical values for normal exposure which may be used as a general guide in computing the heat losses through the structure.

Whether the calculation is carried out by the environmental-temperature concept or the internal air-temperature method, the following formula for the heat loss through the structure is used:

heat loss in watts = area (m^2) × U value × temperature difference

Example 7.7. *Compare the heat losses in watts through two walls having U values of 1.5 W/m^2 K and 0.70 W/m^2 K respectively, when the area of each wall is 200 m^2. The internal environmental temperature may be taken as 20°C when the outside air temperature is − 1°C.*

$$\text{heat loss} = \text{area} \times U \times (t_{ei} - t_{ao})$$

where   area = m^2

U = thermal transmission (W/m^2 K)

t_{ei} = environmental temperature inside ($^{\circ}$C)

t_{ao} = air temperature outside ($^{\circ}$C)

For walls having a U value of 1.50 W/m^2 K,

$$\text{heat loss} = 200 \times 1.50 \times (20 - (-1))$$

$$= 200 \times 1.50 \times 21$$

$$= 6300 \text{ watts}$$

Table 7.4 U values for common types of construction

Construction	U value (W/m² K)
Walls (brickwork)	
220 mm solid brick wall unplastered	2.3
220 mm solid brick wall with 16 mm plaster	2.1
260 mm unventilated cavity wall with 105 mm brick outer and inner leaves, with 16 mm plaster	1.5
260 mm unventilated cavity wall with 105 mm brick outer leaf and 100 mm lightweight concrete block inner leaf and 16 mm plaster	0.96
260 mm cavity wall as above but with 13 mm expanded polystyrene board in cavity	0.70
Flat roofs	
19 mm asphalt on 150 mm solid concrete	3.4
19 mm asphalt on 150 mm hollow tiles	2.2
19 mm asphalt on 13 mm cement and sand screed, 50 mm woodwork slabs on timber joists and aluminium foil backed by 10 mm plasterboard ceiling	0.9
Doors	
25 mm wood or cored	2.4
Partition walls	
75 mm breeze blocks plastered on both sides	2.2
Pitched roofs	
Tiles on battens, sarking felt and plasterboard ceiling	2.44
Tiles on battens, sarking felt and aluminium foil-backed 10 mm plasterboard ceiling	1.50
As above but with 100 mm glass-fibre insulation laid between joists	0.35
Floors	
Solid ground floor with exposed edges 3 m by 3 m	1.47
Solid ground floor with exposed edges 30 m by 30 m	0.26
Solid ground floor with exposed edges 60 m by 60 m	0.11
Intermediate floor on joists with plaster ceiling	1.65
Intermediate floor with 150 mm concrete, 50 mm screed screed and timber flooring	1.87
Suspended timber floor above ground 3 m by 3 m	1.05
Suspended timber floor above ground 30 m by 30 m	0.28
Suspended timber floor above ground 60 m by 60 m	0.16
Glass	
Single glazing	5.6
Double glazing with 20 mm air space	2.9
Triple glazing with 20 mm air space	2.0

For walls having a U value of 0.70 W/m^2 K,

$$\text{heat loss} = 200 \times 0.70 \times 21$$
$$= 2940 \text{ watts}$$

$$\begin{array}{l} \text{difference} \\ \text{in heat loss} \end{array} = 6300 - 2940 = 3360 \text{ watts}$$

$$\begin{array}{l} \text{percentage} \\ \text{saving} \end{array} = \frac{3360}{6300} \times \frac{100}{1}$$

$$= 53.333 \text{ per cent}$$

Heat loss by ventilation

The heat lost due to infiltration of external air may be found from the following formula:

$$Q_v = \rho\, V C\, (t_{ai} - t_{ao})$$

where

Q_v = heat lost (W)

ρ = density of air which may be taken as 1.2 kg/m^3

V = infiltration rate (m^3/s)

C = specific heat capacity of air, which may be taken as 1000 J/kg K

t_{ai} = inside air temperature ($^\circ$C)

t_{ao} = outside air temperature ($^\circ$C)

By introducing the infiltration rate N (air changes per hour) and room volume V (m^3) the formula may be written:

$$Q_v = \frac{\rho\, N V C}{3600}\, (t_{ai} - t_{ao})$$

and substituting the above values for ρ and C,

$$Q_v = \frac{1.2 \times N \times V \times 1000}{3600}\, (t_{ai} - t_{ao})$$
$$= 0.333\, N V\, (t_{ai} - t_{ao})$$

where

Q_v = heat lost in watts

N = rate of air change per hour

V = volume of room m^3

t_{ai} = inside air temperature $^\circ$C

t_{ao} = outside air temperature $^\circ$C

Example 7.8. *Calculate the heat lost by ventilation in a room measuring 10 m by 8 m by 3 m when the rate of air change is three per hour and the inside and outside air temperatures are 22 $^\circ$C and -1 $^\circ$C respectively.*

$$Q_v = 0.333\, N V\, (t_{ai} - t_{ao})$$
$$= 0.333 \times 3 \times 240 \times (22 - (-1))$$
$$= 5514.48 \text{ watts}$$

Example 7.9. *A room measuring 8 m by 6 m by 3 m has a heat input of 6 kW when the air temperatures inside and outside are 20°C and −2°C respectively. If air from the outside passes through a convector, calculate the rate of air change per hour.*

$$Q_v = 0.333 \; N \; V \; (t_{ai} - t_{ao})$$

$$N = \frac{Qv}{0.333 \; V \; (t_{ai} - t_{ao})}$$

$$= \frac{6000}{0.333 \times 144 \times 22}$$

$$= 5.687$$

$$= 6 \text{ air changes per hour (approx.)}$$

Note: Some examination questions give the heat capacity of air in J/m^3 K. A common value given is 1340 J/m^3 K.

The formula for the heat lost by ventilation may be written as follows:

$$\text{watts} = \frac{HC \times \text{volume of room } (m^3) \times \text{air change per hr} \times (t_{ai} - t_{ao})}{3600}$$

where HC = heat capacity of air

Example 7.10. *Figure 7.6 shows the plan of a two-bedroomed bungalow. Using the following factors, calculate the heat losses in kW and the required radiator areas in m^2. The bungalow is to be well insulated and the internal air-temperature method of calculating the heat loss through the external walls may therefore be used. External air temperature −1°C, radiator heat emission 530 W/m^2. Size of doors 1.8 × 0.76.*

Factors

1. Design temperature (°C)		Air change per hr	2. Window areas (m^2)	
Dining room	22	2	Dining room	2.0 × 1
Lounge	22	2	Lounge	1.7 × 1
Kitchen	18	2	Kitchen	1.2 × 1
Bedrooms	18	1½	Bathroom	1.2 × 1
Bathroom	22	2	Bedroom one (2)	1.7 × 1
Hall	18	1½	Bedroom two	1.7 × 1
Corridor	18	1½		

3. U values (W/m^2 K)		4. Air volumes			m^3
Floors	0.30	Dining room	3.0 × 3.0 × 2.4		21.6
External walls	0.60	Lounge	4.3 × 4.1 × 2.4		42.3
Partition walls	2.20	Hall	3.2 × 1.5 × 2.4		11.5
Windows	2.80	Corridor	4.8 × 1.0 × 2.4		11.5
Doors	2.40	Kitchen	3.3 × 3.0 × 2.4		23.7
Ceilings	0.40	Bathroom	3.0 × 2.8 × 2.4		20.1
		Bedroom one	5.0 × 3.0 × 2.4		36.0
		Bedroom two	3.3 × 3.09 × 2.4		24.4

Fig. 7.6 Plan of two-bedroomed bungalow

127

Building Regulations 1985 (Conservation of Fuel and Power)

Exposed elements means in relation to a dwelling:

(a) exposed to the outside air; or

(b) separating a dwelling from part of a building which is ventilated by means of an opening or duct to the outside air, the aggregate area of which exceeds 5 per cent of the area of the walls enclosing the space, and which allows a passage of air at all times

in relation to a building not consisting of a dwelling

(c) exposed to outside air; or

(d) separating a part of the building which is heated from a part which is not and which is wholly or partly exposed to the outside air.

The area of perimeter wall used to calculate the areas of windows should include all openings in the wall; and

(a) an external door with 2 m^2 or more of glazed area should be counted as part of the window area; and

(b) any part of a roof which has a pitch of 70° or more should be counted as walling; and

(c) any opening in a wall other than a window opening and a meter cupboard recess may be counted as part of the wall area; and

(d) lintels, jambs and sills associated with windows and roof lights may be counted as part of the window and roof light area; and

(e) areas of walls, floors and roofs should be between finished internal faces of the building and in the case of a roof, in the plane of the insulation.

Example 7.11 *A detached house has an area of exposed walls of 200 m^2 and the area of windows is 50 m^2. The walls and ceiling have U values of 0.5 W/m^2 K and 0.4 W/m^2 K respectively.*

If the area of the ceiling is 84 m^2 and double glazing is to be used, calculate if the levels of insulation meet the requirements of the Regulations.

area of exposed walls including windows	= 200 m^2
area of exposed walls to be insulated = 96 − area of windows = 200 − 50	= 150 m^2
area of roof to be insulated	= 84 m^2

From Table 7.2 the *U* value of exposed walls and roof are 0.6 W/m^2 K and 0.35 W/m^2 K respectively.

allowable rate of heat loss through exposed walls (150 × 0.6)	= 90 W/K
allowable rate of heat loss through roof (84 × 0.35)	= 29.4 W/K
total allowable rate of heat loss through walls and roof	119.4 W/K

In proposed house

rate of heat loss through exposed walls = (150 × 0.5)		= 75.0
rate of heat loss through roof (84 × 0.4)		= 33.6
total heat loss through walls and roof		108.6 W/K

(The proposed insulation levels are satisfactory)

Note: A good deal of heat loss through window openings may be saved if the windows having the larger areas have a southern aspect.

In winter, south-facing windows during daytime may provide a useful degree of solar heating and buildings are sometimes constructed having most of the windows facing south.

This type of solar heating is known as 'passive' and another type, which circulates water through a solar collector, is known as 'active'. Both types of solar heating are dealt with in detail in Chapter 11.

In order to calculate the heat loss through the structure a suitable table may be used as in Tables 7.5–7.12.

Table 7.5 Dining room

Type of structure	Dimensions (m)	Area (m²)	U value (W/m² K)	Temperature difference (K)	Heat loss (W)
External wall	9 × 2.4 deduct 2 × 1.8 (window)	18	0.6	23	248.40
Window	2 × 1.8	3.6	2.8	23	231.84
Partition wall	3 × 2.4	7.2	2.2	4	63.36
Floor	3 × 3	9	0.3	23	62.10
Ceiling	3 × 3	9	0.4	23	62.10
	heat loss through structure			Total	729.90

Dining room

ventilation losses

$$Q_v = 0.333 \, N \, V \, (t_{ai} - t_{ao})$$
$$= 0.333 \times 2 \times 21.6 \times 23$$
$$= 330.8688$$

total heat losses = 667.80 + 330.87
$$= 998.67 \text{ watts}$$

area of radiator = $\dfrac{998.67}{530}$

$$= 2 \text{ m}^2 \text{ (approx.)}$$

Table 7.6 Lounge

Type of structure	Dimensions (m)	Area (m²)	U value (W/m² K)	Temperature difference (K)	Heat loss (W)
External wall	8.4 × 2.4 deduct 1.7 × 1.3	17.95	8.60	23	241.71
Window	1.7 × 1.3	2.21	2.80	23	142.324
Partition wall	4.1 × 2.4 deduct 1.8 × 0.76	8.472	2.20	4	74.554
Door	1.8 × 0.76	1.368	2.40	4	13.133
Floor	4.3 × 4.1	17.63	0.30	23	121.647
Ceiling	4.3 × 4.1	17.63	0.30	23	121.647
	heat loss through structure			Total	721.015

Lounge

ventilation losses

$$Q_v = 0.333 \, N \, V \, (t_{ai} - t_{ao})$$
$$= 0.333 \times 2 \times 42.312 \times 23$$
$$= 648.135$$

$$\text{total heat losses} = 721.015 + 648.135$$
$$= 1369.150 \text{ watts}$$

$$\text{area of radiator} = \frac{1369.150}{530}$$
$$= 2.6 \text{ m}^2 \text{ (approx.)}$$

Table 7.7 Hall and corridor (heat loss)

Type of structure	Dimensions (m)	Area (m²)	U value (W/m² K)	Temperature difference (K)	Heat loss (W)
External wall	1.5 × 2.4 deduct 1.8 × 0.76	2.232	0.70	19	25.50
Door	1.8 × 0.76	1.368	2.40	19	62.380
Floor	(4.1 × 1.5) add	9.45	0.30	19	53.865
Ceiling	3.3 × 1.0	2.85	0.40	19	16.245
	heat loss through structure			Total	157.990

Hall and corridor

In order to find the effective heat loss through the structure, the heat gains must be deducted from the heat losses.

Table 7.8 Hall and corridor (heat gains)

Type of structure	Dimensions (m)	Area (m²)	U value (W/m² K)	Temperature difference (K)	Heat loss (W)
Partition	(2.8 + 4.1) × 2.4 deduct 2 doors 1.8 × 0.76	13.824	2.20	4	121.651
Doors	2 × 1.8 × 0.76	2.736	2.40	4	26.265
	heat gains through structure			Total	147.916

effective heat loss = 157.990 − 147.916

$$= 10.074 \text{ watts}$$

ventilation losses

$$Q_v = 0.333\, N\, V\, (t_{ai} - t_{ao})$$
$$= 0.333 \times 1.5 \times (2 \times 11.52) \times (18 - (-1))$$
$$= 0.333 \times 1.5 \times 23.04 \times 19$$
$$= 218.66$$

total heat losses = 10.074 + 218.66

$$= 228.734 \text{ watts}$$

$$\text{area of radiator} = \frac{228.734}{530}$$
$$= 0.432$$
$$= 0.5 \text{ m}^2 \text{ (approx.)}$$

Table 7.9 Bathroom (heat losses)

Type of structure	Dimensions (m)	Area (m²)	U values (W/m² K)	Temperature difference (K)	Heat loss (W)
External wall	2.8 × 2.4 deduct 1.2 × 1.0	5.52	0.60	23	76.176
Window	1.2 × 1.0	1.2	2.80	23	77.280
Partition walls	11.6 × 2.4 deduct 7.8 × 0.76	26.472	2.20	4	232.954
Door	1.8 × 0.76	1.368	2.40	4	13.133
Floor	3.0 × 2.8	8.4	0.30	23	57.96
Ceiling	3.0 × 2.8	8.4	0.30	23	57.96
	heat loss through structure			Total	515.463

Bathroom

ventilation losses

$$Q_v = 0.333\, N\, V\, (t_{ai} - t_{ao})$$

131

$$= 0.333 \times 2 \times 20.16 \times 23$$

$$= 308.80$$

total heat losses $= 515.463 + 308.8$

$$= 824.263 \text{ watts}$$

area of radiator $= \dfrac{824.263}{530}$

$$= 1.55$$

$$= 1.6 \text{ m}^2 \text{ (approx.)}$$

Table 7.10   Bedroom 1

Type of structure	Dimensions (m)	Area (m²)	U value (W/m² K)	Temperature difference (K)	Heat loss (W)
External walls	11 × 2.4 deduct 2 (1.7 × 1.2)	22.32	0.70	19	254.50
Windows	2 (1.7 × 1.2)	4.08	2.80	19	217.056
Floor	5 × 3	15.0	0.40	19	85.500
Ceiling	5 × 3	15.0	0.30	19	85.50
					642.556
Heat gain					
Partition wall	3 × 2.4	7.2	2.2	4	63.360
	heat loss through structure			Total	579.196

Bedroom 1

ventilation losses

$$Q_v = 0.333 \, N \, V \, (t_{ai} - t_{ao})$$

$$= 0.333 \times 1.5 \times 36 \times 19$$

$$= 341.658$$

total heat losses $= 579.196 + 341.658$

$$= 920.854 \text{ watts}$$

area of radiators $= \dfrac{920.854}{530}$

$$= 1.737$$

Use two radiators, 1 m² each.

Kitchen

ventilation losses

$$Q_v = 0.333 \, N \, V \, (t_{ai} - t_{ao})$$

$$= 0.333 \times 2 \times 23.76 \times 19$$

$$= 300.659$$

total heat loss $= 300.659 - 3.598$

$$= 297.061 \text{ watts}$$

132

The heat losses from the boiler would adequately cover the above heat loss and therefore a radiator in the kitchen is not normally required. The heat loss from the kitchen, however, must be added to the heat losses from the other rooms in order to find the boiler power.

Table 7.11 Kitchen

Type of structure	Dimensions (m)	Area (m²)	U value (W/m² K)	Temperature difference (K)	Heat loss (W)
External walls	3.3 × 2.4 deduct 1.2 × 1.0 + 1.8 × 0.76	5.352	0.60	19	61.00
Window	1.2 × 1.0	1.20	2.80	19	6.384
Door	1.8 × 0.76	1.368	2.40	19	6.238
Floor	3.3 × 3.0	9.9	0.30	19	56.430
Ceiling	3.3 × 3.0	9.9	0.30	19	56.43
					186.482

Heat gain

Type of structure	Dimensions (m)	Area (m²)	U value (W/m² K)	Temperature difference (K)	Heat loss (W)
Partition walls	9.0 × 2.4	21.6	2.20	4	190.080
	heat gain through structure			Total	3.598

Table 7.12 Bedroom 2

Type of structure	Dimensions (m)	Area (m²)	U value (W/m² K)	Temperature difference (K)	Heat loss (W)
External walls	(3.3 + 1.9) × 2.4 deduct 1.7 × 1.2	10.44	0.6	19	119.00
Window	1.7 × 1.2	2.04	2.8	19	108.528
Floor	3.3 × 3.09	10.197	0.3	19	58.123
Ceiling	3.3 × 3.09	10.197	0.3	19	58.123
	heat loss through structure			Total	343.774

Bedroom 2
ventilation losses

$$Q_v = 0.333 \, N \, V \, (t_{ai} - t_{ao})$$
$$= 0.333 \times 1.5 \times 24.47 \times 19$$
$$= 232.233$$

$$\text{total heat losses} = 343.774 + 232.233$$
$$= 576.007 \text{ watts}$$

$$\text{area of radiator} = \frac{576.007}{530}$$
$$= 1.086$$
$$= 1.1 \text{ m}^2 \text{ (approx.)}$$

133

Total heat losses (watts)

Dining room	998.670
Lounge	1369.150
Hall and corridor	228.734
Bathroom	824.263
Bedroom 1	920.854
Bedroom 2	576.007
Kitchen	297.061
	5214.739

Boiler power

To this heat loss, an allowance of 20 per cent may be added for the heat losses from the pipes and an allowance of 3000 watts for the hot-water supply.

$$\text{boiler power} = 5215 + 1043 + 3000$$
$$\text{boiler power} = 9258$$
$$= 10 \text{ kW (approx.)}$$

Convective heating using the environmental-temperature concept

The following steps are taken:

1. Calculate the heat loss through the various elements of the structure using the difference between the inside environmental temperature and the outside air temperature and sum these to produce ΣQf.
2. Calculate the area of the entire enclosure ΣA (this should include any partitions, floors, or ceilings gaining heat from or losing heat to the adjacent rooms).
3. Calculate $\Sigma Qf/\Sigma A$ and from the following equation find the difference between the inside air temperature and the outside environmental temperature:

$$t_{ai} - t_{ei} = \Sigma Qf/4.8 \, \Sigma A$$

where t_{ai} = inside air temperature $^\circ$C

t_{ei} = inside environmental temperature $^\circ$C

ΣQf = rate of heat transfer through the building fabric W

ΣA = total area of room surfaces in m^2

4. Calculate the heat loss by ventilation using the formula given previously:

$$Q_v = 0.333 \, N \, V \, (t_{ai} - t_{ao})$$

5. Add the results of steps (1) and (4) to give the total heat loss.

Example 7.12 *Using the bungalow given in Fig. 7.6 and the same U values as given in example 7.10, calculate the rate of heat loss and the surface area of the radiator required for the dining room when the internal environmental temperature is to be 21 $^\circ$C and convective heating is to be used.*

Table 7.13 Dining room

Type of structure	Dimensions (m)	Area (m^2)	U value (W/m^2 K)	Temperature difference (K)	Heat loss (W)
External wall	9 × 2.4 deduct 2 × 1.8 (window)	18.0	0.6	22	237.60
Window	2 × 1.8	3.6	2.8	22	221.76
Partition wall	3 × 2.4	7.2	2.2	3	47.52
Floor	3 × 3	9.0	0.3	22	59.40
Ceiling	3 × 3	9.0	0.3	22	59.40
	ΣA	46.8		ΣQf	685.08

internal air temperature

$$t_{ai} - t_{ei} = \frac{\Sigma Qf}{4.8\,\Sigma A}$$

$$t_{ai} - t_{ei} = \frac{625.68}{4.8 \times 46.8}$$

$$t_{ai} - t_{ei} = 3$$

$$t_{ai} = 21 + 3$$

$$= 24\,^{\circ}\text{C (approx.)}$$

heat loss by ventilation

$$Q_v = 0.333\,N\,V\,(t_{ai} - t_{ao})$$

$$= 0.333 \times 2 \times 21.6 \times 25$$

$$= 359.64$$

$$\text{total heat losses} = 685.08 + 359.64$$

$$= 1044.72 \text{ watts}$$

$$\text{area of radiator} = \frac{1044.72}{530} = 1.97$$

$$= 2\ \text{m}^2 \text{ (approx.)}$$

Radiant heating using the environmental-temperature concept

In order to find the total heat losses the following steps are taken:

1. Calculate the heat loss through the structure as for convective heating to produce ΣQf.
2. Calculate the surface area of the entire enclosure as for convective heating to produce ΣA.

3. Find the ventilation conduction C_v from:

$$C_v = \frac{1}{0.333 \, N \, V} + \frac{1}{4.8 \, \Sigma A}$$

where N = number of air changes per hour

V = volume of the enclosure m^2

4. Calculate ventilation loss Q_v from:

$$Q_v = C_v \, (t_{ei} - t_{ao})$$

5. Add the results of steps 1 and 4 to give the total heat loss.

Example 7.12. *Using the bungalow given in Fig. 7.6 and the same U values given in example 7.10, calculate the rate of heat loss and the surface area of a radiant panel required for the dining room when the internal environmental temperature is to be 21 °C.*

$$\Sigma Qf = 685.08 \text{ W}$$
$$\Sigma A = 46.8 \text{ m}^2$$
$$N = 2 \text{ per hour}$$
$$V = 21.6 \text{ m}^3$$

ventilation conductance

$$\frac{1}{C_v} = \frac{1}{0.333 \, N \, V} + \frac{1}{4.8 \, \Sigma A}$$

$$= \frac{1}{0.333 \times 2 \times 21.6} + \frac{1}{4.8 \times 26.8}$$

$$= 0.07 + 0.008$$

$$C_v = 12.82 \text{ W K}$$

heat loss by ventilation

$$Q_v = C_v \, (t_{ei} - t_{ao})$$
$$= 12.82 \, (21 - (-1))$$
$$= 12.82 \times 22$$
$$= 282.04 \text{ W}$$

total heat loss

$$Q_t = \Sigma Qf + \Sigma Q_v$$
$$= 685.08 + 282.04$$
$$= 967.12$$

area of radiant panel
$$= \frac{967.12}{530}$$
$$= 1.8 \text{ m}^2$$

Note: The heat required for convective heating is higher than the heat required for radiant heating. The human body loses heat by radiation, convection and

136

evaporation by the following approximate percentages:

radiation   = 45 per cent
convection   = 30 per cent
evaporation = 25 per cent

Since the greatest heat lost is by radiation, the provision of radiant heating will 'balance' this loss and will provide thermal comfort for the occupants of the room at a lower heat input.

Although the term radiant heating is used, convective heating is still provided by the radiant panels which should be sufficient to 'balance' the heat lost by convection.

The heat lost by evaporation can only be controlled by: (a) correct relative humidity of the air which is usually between 40 per cent and 60 per cent, and an air-conditioning system would be required for this purpose; (b) by the correct air velocity which is about 0.2 m/s.

Heat emission from a radiator

Manufacturers usually give the radiator heat emission for a temperature differential between the mean water temperature and the ambient, or surrounding, air temperature of 55.6 °C.

For a temperature differential of other than 55.6 °C, the radiator heat emission may be found from the following formula:

$$E_2 = E_1 \times \left[\frac{\Delta t}{55.6}\right]^{1.3}$$

where E_2 = heat emission in W/m^2 for the new temperature differential

E_1 = heat emission in W/m^2 for a 55.6 °C differential

Δt = the new temperature differential in °C

Example 7.13. *Calculate the heat emission from a radiator in W/m^2 when the temperatures of the flow and return waters are 80 °C and 70 °C respectively, when the ambient air temperature is 18 °C. The heat emission for the radiator given by the manufacturers for a temperature differential of 55.6 °C is 520 W/m^2.*

$$\Delta t = \left[\frac{80 + 70}{2}\right] - 18$$

$$\Delta = 75 - 18$$
$$= 57 °C$$

Substituting in the formula,

$$E_2 = 520 \times \left[\frac{57}{55.6}\right]^{1.3}$$

$$= 520 \times 1.025^{1.3}$$

$$= 520 \times 1.030$$

$$= 535.6 \text{ W/m}^2$$

Heat emission from a convector

When considering the heat emission from natural convectors for temperature differentials other than 55.6 °C, the following formula may be used:

$$E_2 = E_1 \times \left[\frac{\Delta t}{55.6} \right]^{1.5}$$

If the heat emitter in example 7.13 is a natural convector, the heat emission would be:

$$E_2 = 520 \left[\frac{57}{55.6} \right]^{1.5}$$
$$= 539.76 \text{ W/m}^2$$

Questions

1. Define the following terms: (a) thermal conductivity; (b) thermal resistivity; (c) thermal conductance; (d) thermal resistance; (e) thermal transmission.

2. One wall of a building consists of 25 mm thickness of cedar boardings; 76 mm thickness of glass wool; and 13 mm thickness of plasterboard. Using the following values, calculate the thermal transmission (U) for the wall.

Thermal conductivity

Cedar wood	0.14 W/m K
Glass wool	0.042 W/m K
Plasterboard	0.58 W/m K

Thermal resistances

Inside surface	0.123 m² K/W
Outside surface	0.055 m² K/W

Answer: 0.457 W/m² K

Fig. 7.7 Plan of an office

3. If the thermal transmission for a wall is 0.74 W/m² K, calculate the heat loss in watts through the wall having an area of 36 m².

Answer: 2.664 watts

4. Figure 7.7 shows the plan of an office having an internal environmental temperature of 20 °C when the external air temperature is −1 °C. Using the environmental concept for convective heating and an air change of 2 per hour, calculate the rating of the convector heaters in kW. Use the following values:

    (a) area of each window 4 m².
    (b) area of each door 1.4 m².

    (c) U values (W/m² K)

Floor	1.40
External walls	0.96
Partition walls	2.2
Windows	5.7
Doors	2.4
Ceiling	0.9

Answer: 17.4 kW (approx.). Use three heaters, 5.8 kW each.

5. Using the following data, calculate the environmental temperature of a room in which the air temperature is 22 °C:

Fabric	Area (m²)	Surface temperature (°C)
Floor	150	20
Walls	150	20
Windows	30	8
Ceiling	150	18

Answer: 19.73 °C

6. A pitched roof inclined at 30° has a horizontal ceiling below. If the U values for the roof and ceiling are found to be 2.80 W/m² K and 0.7 W/m² K respectively, calculate the combined U value for the ceiling and roof. Does this value satisfy the Building Regulations for offices?

Answers: 0.575 W/m² K; yes, it is below 0.6

7. Calculate the heat emission from a radiator in W/m² when the temperatures of the flow and return waters are 82 °C and 71 °C respectively, when the ambient air temperature is 16 °C. The heat emission for the radiator given by the manufacturers for a temperature differential of 55.6 °C is 530 W/m².

Answer: 591.5 W/m² (approx.)

8. Calculate the heat lost by ventilation in a room measuring 15 m by 10 m by 3 m, when the rate of air change is six per hour and the inside and outside air temperatures are 24 °C and −10 °C respectively.

Answer: 30.569 kW (approx.)

Chapter 8

Pipe sizing for heating, pump duty

Pumped circulation

Pumped circulation is used for all medium and large installations and has the following advantages over natural or gravity circulation:

1. Quicker response to heating of the building.
2. Circulation is independent of the temperature difference between the flow and return water.
3. Smaller pipes may be used, thus saving in capital costs, saving in space, and neater in appearance.
4. The boiler room may be sited on the roof.
5. Any type of heat emitter may be used.

Although the cost of the pump or pumps and the cost of power for running them must be considered, the use of pumped circulation adequately compensates by the saving in the overall capital and running cost of the installation.

The following steps may be taken in finding the sizes of the heating pipes:

1. Calculate the heat requirements for each room as described in Chapter 7.
2. Make a sketch of the system showing the positions of the heat emitters, boiler and the pipe runs. At this stage it will be necessary to decide on whether the system is to be one- or two-pipe and also the number of circuits to serve the various rooms in which heat emitters are to be installed.
3. Decide on the temperature of the flow and return water, i.e. the temperadrop. Although this may appear arbitrary, the diameter of the pipe and the pump duty found later will provide for this temperature drop.

4. Estimate the heat emission to be expected from the pipes. It is usual to allow between 5 and 30 per cent of the total heat from the emitters for the heat emission from the pipes. The amount will depend upon the diameter and length of pipes.
5. Estimate the mass rates of flow through the circuit or circuits in kg/s.
6. Decide upon the velocity of water flow.
7. From the pipe-sizing table find a suitable size pipe or pipes.
8. Measure the actual lengths of the circuit or circuits and to this add an allowance for frictional losses due to fittings, to obtain the effective pipe length.
9. Find the pressure required for the pump.

An example will show how these steps are carried out.

Example 8.1. *Figure 8.1 shows a single-pipe circuit. Assuming each radiator has a heat emission of 8 kW and the temperature drop across the system is 15 °C, find by the use of Table 8.1 the approximate diameter of a steel pipe for the main and the pump pressure. Use a velocity of flow of water of 1 m/s.*

Fig. 8.1 Single-pipe ring system

The single-pipe circuit must carry sufficient energy in the water flowing through it to supply the total heat required for the radiators, plus the heat required for the pipes.

total heat emission from radiators = 48 kW
heat emission from the pipes
assuming 20 per cent of the heat
emission from radiators   = 9.6 kW
                               total   = 57.6

141

Mass flow rate (kg/s)

$$\text{power} = \text{kg/s} \times \text{s.h.c.} \times (t_f - t_r) \quad \text{(kW)}$$

$$\therefore \quad \text{flow rate} = \frac{\text{kW}}{\text{s.h.c.} \times (t_f - t_r)} \quad \text{(kg/s)}$$

where kW = total heat carried by the pipe

s.h.c. = specific heat capacity of water taken as
4.2 kJ/kg °C

t_f = temperature of flow water °C

t_r = temperature of return water °C

$$\text{flow rate} = \frac{57.6}{4.2 \times 15}$$

$$= 0.92 \text{ kg/s}$$

Table 8.1 Flow of water at 75°C in black steel pipes. Medium-grade pipe to
BS 1387:1967

	20 mm		25 mm		32 mm		40 mm		50 mm			Pressure drop (Pa per m
V	M	EL	M	EL	M	EL	M	EL	M	EL	V	run of pipe)
	0.122	0.7	0.228	1.0	0.480	1.4	0.720	1.8	1.35	2.4		80.0
	0.124	0.7	0.232	1.0	0.488	1.4	0.732	1.8	1.37	2.4		82.5
	0.126	0.7	0.236	1.0	0.496	1.4	0.743	1.3	1.39	2.4		85.0
	0.128	0.7	0.240	1.0	0.503	1.4	0.755	1.8	1.41	2.4		87.5
	0.130	0.7	0.243	1.0	0.511	1.4	0.766	1.8	1.43	2.4		90.0
	0.132	0.7	0.247	1.0	0.518	1.5	0.778	1.8	1.45	2.4		92.5
	0.134	0.7	0.251	1.0	0.526	1.5	0.789	1.8	1.48	2.4		95.0
	0.136	0.7	0.254	1.0	0.533	1.5	0.800	1.8	1.50	2.4		97.5
	0.138	0.7	0.258	1.0	0.540	1.5	0.810	1.8	1.52	2.4		100.0
	0.152	0.7	0.284	1.0	0.595	1.5	0.893	1.8	1.67	2.4		120.0
	0.165	0.8	0.308	1.0	0.646	1.5	0.968	1.8	1.81	2.5		140.0
0.5	0.178	0.8	0.331	1.0	0.693	1.5	1.04	1.8	1.94	2.5		160.0
	0.189	0.8	0.353	1.0	0.738	1.5	1.11	1.8	2.06	2.5	1.0	180.0
	0.200	0.8	0.373	1.0	0.780	1.5	1.17	1.9	2.18	2.5		200.0
	0.211	0.8	0.392	1.1	0.820	1.5	1.28	1.9	2.29	2.5		220.0
	0.221	0.8	0.411	1.1	0.858	1.5	1.29	1.9	2.40	2.5		240.0
	0.230	0.8	0.428	1.1	0.895	1.5	1.34	1.9	2.50	2.5		260.0
	0.239	0.8	0.445	1.1	0.931	1.5	1.39	1.9	2.60	2.6		280.0
	0.248	0.8	0.462	1.1	0.965	1.5	1.44	1.9	2.69	2.6		300.0
1.0	0.257	0.8	0.478	1.1	0.998	1.6	1.49	1.9	2.78	2.6	1.5	320.0

V = velocity of flow m/s
M = mass rate of flow kg/s
EL = equivalent length factor

From the pipe-sizing table (Table 8.1), a 40 mm diameter steel pipe will carry
0.968 kg/s at a velocity of flow of 1 m/s giving a pressure loss per metre run
of 140 Pa and an EL factor of 1.8.

Check on the estimated heat loss on the pipes of 20 per cent

Table 8.2 shows that for a 50°C difference between the pipe surface and the
surrounding air, the heat loss on a 40 mm diameter pipe is 105 watts per
metre run.

Table 8.2 Theoretical heat emission from single horizontal steel pipes freely exposed in ambient air

Nominal bore (mm)	Temperature difference of surface to surroundings (°C)				
	40	45	50	55	60
8	28	32	35	39	43
10	33	37	41	46	50
15	40	46	53	59	66
20	48	56	62	70	78
25	58	68	78	88	98
32	71	82	93	110	120
40	78	92	105	110	130
50	96	112	130	150	170
65	120	140	160	180	200
80	140	160	180	211	230

Heat emission (W/m run)

Since the total length of pipe is 82 m,

heat emission from pipe = 82 × 105 = 8610 watts

$$\text{percentage emission} = \frac{8610}{48\,000} \times \frac{100}{1} = 18 \text{ per cent}$$

This is only a 2 per cent difference between the estimated heat loss from the pipes and the actual heat loss, and therefore the pipe diameter is suitable.

Pump duty

Before the pressure required by the pump can be determined, it will be necessary to find the effective length of the pipe. The effective length is the addition of the equivalent length due to the resistances of fittings and the actual length.

The resistances of the fittings are computed, in terms of equivalent length of straight pipe of corresponding diameter, by the product of an equivalent length factor found from Table 8.1 and a velocity pressure-loss factor, K.

Table 8.3 gives the K factor for some of the common types of fittings.

The pressure created by a pump on a one-pipe system does not force water through the radiators (unless special injector tees are used). The hot water therefore circulates through the radiators by natural convection created by the difference in density of the water flowing into the radiators and the cooler, denser water flowing out.

The frictional resistances through the tees for the purpose of determining the effective length of pipe will therefore be ignored.

Equivalent length of pipe = number of fittings × K × EL.

From Table 8.3, the K factor for a boiler is 5; and for 40 mm diameter

Table 8.3 Values of velocity pressure factor K for pipe fittings and equipment

Tees	K factor
	0.5 plus 0.2 for enlargement or reduction
Gate valve	0.2
Angle valve	0.5
Radiator	5.0
Sectional boiler	5.0
$90°$ elbow 10 mm to 25 mm	0.8
$90°$ elbow 32 mm to 50 mm	0.7
$90°$ elbow 65 mm to 90 mm	0.6
$90°$ bend 10 mm to 25 mm	0.7
$90°$ bend 32mm to 50 mm	0.5
$90°$ bend 65 mm to 90 mm	0.4
Reductions or enlargements on a straight run of pipe	
3 : 2 reduction	0.3
2 : 1 reduction	0.4
3 : 1 reduction	0.4
4 : 1 reduction	0.5
3 : 2 enlargement	0.4
2 : 1 enlargement	0.7
3 : 1 enlargement	0.9

$90°$ bends, it is 0.5. The equivalent lengths are therefore:

$$boiler = 1 \times 5 \times 1.8 = 9.0 \text{ m}$$
$$bends = 6 \times 0.5 \times 1.8 = 5.4 \text{ m}$$
$$14.4 \text{ m}$$

$$actual \ length = 1 + 12 + 25 + 15 + 25 + 2 + 2 = 82 \text{ m}$$
$$\therefore \ effective \ length \ of \ pipe = 82 + 14.4 = 96.4 \text{ m}$$

Pump pressure

The pressure loss through the pipe from Table 8.1 was found to be 140 Pa per metre run of pipe.

$$\therefore \ total \ pressure \ loss = 96.4 \times 140 = 13\ 496 \text{ Pa}$$

This is approximately equal to 13.5 kPa pressure and is equal to
$13.5 \div 9.81 = 1.4$ m head of water.

Pumps are ordered giving the pressure developed and the flow rate in kg/s or litre/s. A pump providing a pressure of 13.5 kPa and giving a mass flow rate of 1 kg/s would be satisfactory.

Two-pipe system

In a two-pipe system, the flow pipes will reduce in diameter as they pass along the circuit, and the return pipes will increase in diameter as they return to the boiler.

The pressure created by a pump in a two-pipe system acts through the heat emitters and therefore convector heaters, which offer more resistance to the flow of water than radiators, can be used. The frictional resistances of the tees will, however, have to be taken into account when determining the effective length of pipe and the pump pressure.

Index circuit

In order to determine the pump pressure, the index circuit has to be selected. This is the circuit having the greatest resistance to the flow of water and supplies the index heat emitter. For a pumped system the index circuit is always taken to be the longest circuit.

Heat distribution

Unlike the one-pipe system, where the cooler water from the heat emitter passes back into the flow pipe, in the two-pipe system, the cooler water from each heat emitter passes into the return pipe and this provides a better balance of heat distribution in the system.

Example 8.2. *Figure 8.2 shows a two-pipe parallel system to be used as an alternative to the one-pipe ring system given in example 8.1. Using the follow-data and assumptions find by use of Tables 8.1—8.3 the approximate diameters of the flow and return mains and pump duty.*

Fig. 8.2 Two-pipe parallel system

1. *Heat emission from each convector* = *12 kW*
2. *Temperature drop across the system* ≑ *15 °C*
3. *Mean temperature of water* = *72 °C*
4. *Specific heat capacity of water* = *4.2 kJ/kg K*
5. *Velocity of flow of water* = *1 m/s*

Pipe sizing

Before the pipe-sizing procedure is commenced, a reference number or letter
should be allocated to each section of pipework for identification purposes.
The pipe-sizing may be started from the boiler or the farthest convector from
it. In this example, the index convector is used as the starting point.

In order to find the mass flow rate through each section of pipework,
an estimate is made of the carrying capacity of the pipes for each section,
by taking the heat required for the convector and adding an estimated percentage for the pipework heat emission. This estimate can be checked and the pipe-
sizing revised, if required.

Pipes number 1: Heat emission from convector 1 and pipework 1

$$\text{emission from convector} = 12.0 \text{ kW}$$
$$\text{emission from pipework} = 2.4 \text{ kW}$$
$$5 \text{ per cent of } 48 \text{ kW}$$
$$\text{total} = 14.4 \text{ kW}$$

$$\text{flow rate} = \frac{14.4}{4.2 \times 15}$$

$$= 0.228 \text{ kg/s}$$

From pipe-sizing Table 8.1, a 20 mm diameter steel pipe will carry 0.257 kg/s
giving a pressure loss per metre run of 320 Pa and an EL factor of 0.8.

Check on estimated heat loss of 5 per cent assuming a temperature of pipe
surface to the ambient air of 50 °C (see Table 8.2).

$$26 \text{ m} \times 62 = 1612 \text{ watts}$$

$$\text{percentage emission} = \frac{1612}{48\,000} \times \frac{100}{1}$$

$$\text{heat loss} = 3.4 \text{ per cent}$$

Since 5 per cent was allowed, pipes number 1 are satisfactory.

Pipes number 2: Emission from convector 2 and pipework 2

$$\text{emission from convector} = 12.0 \text{ kW}$$
$$\text{emission from pipework} = 4.8 \text{ kW}$$
$$10 \text{ per cent of } 48 \text{ kW}$$
$$\text{total} = 16.8 \text{ kW}$$

$$\text{flow rate} = \frac{16.8 \text{ kW}}{4.2 \times 15}$$

$$= 0.267 \text{ kg/s}$$

Pipes number 2 will also have to carry the heat required for emitter and pipework 1.

$$\text{total flow rate} = 0.228 + 0.267$$

$$= 0.5 \text{ kg/s}$$

From the pipe-sizing Table 8.1, a 32 mm diameter pipe will carry 0.503 kg/s giving a pressure loss per metre run of 87.5 Pa and an EL factor of 1.4.

Check on estimated heat loss of 10 per cent (see Table 8.2).

$$54 \text{ m} \times 93 = 5022 \text{ watts}$$

$$\text{percentage heat loss} = \frac{5022}{48\ 000} \times \frac{100}{1}$$

$$\text{heat loss} = 10.46 \text{ per cent}$$

Pipes number 2 are therefore satisfactory for practical purposes.

Pipes number 3: Emission from convector and pipework 3

$$\text{emission from convector} = 12.0 \text{ kW}$$
$$\text{emission from pipework} = 2.4 \text{ kW}$$
$$5 \text{ per cent of } 48 \text{ kW}$$
$$\text{total} = 14.4 \text{ kW}$$

$$\text{flow rate} = \frac{14.4}{4.2 \times 15}$$

$$= 0.228 \text{ kg/s}$$

Pipes number 3 will also have to carry the heat required for emitters and pipework 1 and 2.

$$\text{total flow rate} = 0.5 + 0.228$$

$$= 0.728 \text{ kg/s}$$

From pipe-sizing Table 8.1, a 32 mm diameter pipe will carry 0.738 kg/s giving a pressure loss per metre run of 180 Pa and an EL factor of 1.5.

Check on estimated heat loss of 5 per cent (see Table 8.2).

$$26 \text{ m} \times 93 = 2418 \text{ watts}$$

$$\text{percentage heat loss} = \frac{2418}{48\ 000} \times \frac{100}{1}$$

$$\text{heat loss} = 5 \text{ per cent}$$

Pipes number 3 are therefore satisfactory.

Pipes number 4: Emission from convector 4 and pipework 4

$$\text{emission from convector} = 12.0 \text{ kW}$$
$$\text{emission from pipework} = 2.4 \text{ kW}$$
$$5 \text{ per cent of } 48 \text{ kW}$$
$$\text{total} = 14.4 \text{ kW}$$

$$\text{flow rate} = \frac{14.4}{4.2 \times 15}$$

$$= 0.228 \text{ kg/s}$$

Pipes number 4 will also have to carry the heat required for emitters and pipework 1, 2 and 3.

$$\text{total flow rate} = 0.728 + 0.228$$

$$= 1 \text{ kg/s}$$

From pipe-sizing Table 8.1, a 40 mm diameter steel pipe will carry 1.04 kg/s giving a pressure loss per metre run of 160 Pa and an EL factor of 1.8.

Check on estimated heat loss of 5 per cent (see Table 8.2).

$$19.5 \text{ m} \times 105 = 2047.5 \text{ watts}$$

$$\text{percentage heat loss} = \frac{2047.5}{48\ 000} \times \frac{100}{1}$$

$$\text{heat loss} = 4.3 \text{ per cent (approx.)}$$

Pipes number 4 are therefore satisfactory for practical purposes;

Pump duty

Before the pressure required by the pump can be determined, it is necessary to find the effective length of each section of pipework in the system. The effective lengths of pipe are given by addition of the equivalent length due to the resistances of the fittings and the actual lengths.

Effective lengths (see Table 8.3)

Pipes number 1, 20 mm diameter with an actual length of 26 m.

bends	$= 2 \times 0.7 \times 0.8 =$	1.12 m
angle valves	$= 2 \times 0.5 \times 0.8 =$	0.80 m
tees	$= 2 \times 0.7 \times 0.8 =$	1.12 m
emitter	$= 1 \times 5.0 \times 0.8 =$	4.00 m
	total =	7.04 m

effective length = 26 + 7.04 = 33.04 m

Pipes number 2, 32 mm diameter with an actual length of 54 m.

bends	$= 2 \times 0.5 \times 1.4 =$	1.40 m
angle valves	$= 2 \times 0.5 \times 1.4 =$	1.40 m
tees	$= 2 \times 0.7 \times 1.4 =$	1.96 m
emitter	$= 1 \times 5.0 \times 1.4 =$	7.00 m
	total =	11.76 m

effective length 54 + 11.76 = 65.76 m

Pipes number 3, 32 mm diameter with an actual length of 26 m.

tees	$= 2 \times 0.7 \times 1.5 =$	2.1 m
angle valves	$= 2 \times 0.5 \times 1.5 =$	1.5 m
emitter	$= 1 \times 5.0 \times 1.5 =$	7.5 m
	total =	11.1 m

effective length = 26 + 11.1 = 37.1 m

Pipes number 4, 40 mm diameter with an actual length of 18.5 m.

bends	$= 3 \times 0.5 \times 1.8 =$	2.70 m
angle valves	$= 2 \times 0.5 \times 1.8 =$	1.80 m
tees	$= 2 \times 0.7 \times 1.8 =$	2.52 m
emitter	$= 1 \times 5.0 \times 1.8 =$	9.00 m

boiler $\quad= 1 \times 5.0 \times 1.8 = \underline{9.00 \text{ m}}$

$\quad\quad\quad\quad\quad\quad$ total $= \underline{25.02 \text{ m}}$

effective length $= 18.5 + 25 \quad = 43.50 \text{ m}$

Pump pressure

The pressure developed by the pump will have to overcome the frictional resistances through all the effective lengths of pipes to the index convector. A table may now be compiled (see Table 8.4).

Table 8.4 Total pressure loss on the system

Pipe number	Effective pipe length (m)	Pressure loss per metre run (Pa)	Pressure loss on section (Pa)
1	33.04	320.0	10 572.8
2	65.76	87.5	5 754.0
3	37.10	180.0	6 678.0
4	43.50	160.0	6 960.0
		Total pressure loss	29 964.8 Pa

A pump providing a pressure of 30 kPa (approx.) and giving a mass flow rate of 1 kg/s would be satisfactory.

Boiler power

The boiler will have to have sufficient power to supply heat required from the pipes and the convectors.

boiler power $= 14.4 + 16.8 + 14.4 + 14.4$

$\quad\quad\quad\quad\quad\quad = 60$ kW

A 10 per cent boiler margin may be added for pre-heating and to give a reserve of power for severe weather conditions.

$\quad$ Therefore the boiler power would be 66 kW.

Micro-pipe systems

Micro-pipe systems can be 'open' or 'sealed' circuits. An open circuit is provided with an expansion and feed cistern with a cold feed and vent pipe, which are open to the atmosphere. A closed or sealed system dispenses with an expansion and feed cistern, and the open cold feed and vent pipes, and an expansion vessel is substituted in their place. This expansion vessel contains a diaphragm with nitrogen or air on one side and water from the heating system on the other. When the water is heated, the expansion is taken up by compressing the nitrogen or air. The Heating and Ventilating Contractors Association in its 'Guide to Good Practice' limits the flow-water temperature in open-circuit micro-pipe systems to 82 °C and in closed or sealed systems to 99 °C. Open circuits are usually operated with a temperature difference between the flow and return water of 11 °C, while sealed systems usually operate at a temperature difference between flow and return water of up to 20 °C. This higher temperature in the sealed systems increases the carrying capacity of the pipes.

The outside diameter of the pipes to the heat emitters may be 6 mm, 8 mm, or 10 mm, depending upon the heating load and the pump pressure.

Circuit design

The following factors are usually taken into account:

1. To prevent air-locking, the velocity of flow of water should be as high as possible; up to 1.5 m/s and above 0.3 m/s.
2. In determining the mass flow rate, an allowance is made for the heat loss from the pipework. This loss is generally taken as 5 per cent for the larger-diameter mains and 10 per cent for the micro-pipe, of the heat emission from the radiators each serve.

The following steps may be taken in finding the sizes of the heating pipes:

1. Calculate the heat requirements for each room as described in Chapter 7.
2. Make a sketch of the system showing the positions of the heat emitters, boiler and pipe runs. The micro-pipe system is always a two-pipe system.
3. Allocate a reference number to each section of pipework.
4. Decide on the temperature difference between the water in the flow and return pipes.
5. Estimate the mass flow rate through each circuit, allowing 5 or 10 per cent heat losses on the pipes.
6. Allowing a velocity of flow of water above 0.3 m/s and up to 1.5 m/s, find a suitable diameter pipe from Tables 8.5 and 8.6.
7. Measure the actual lengths of the circuits and to these add an allowance for frictional losses due to fittings to obtain the effective pipe length.
8. Find the 'index run' and determine the pressure drop along it in order to ascertain the pump pressure.

An example will show how these steps are carried out.

Example 8.3. *Figure 8.3 shows a micro-pipe system for a three-bedroomed house. If an 'open' type of system is to be used with flow and return water temperatures of 80 °C and 69 °C respectively, find the approximate diameters of the pipes and the pump duty.*

Pipes number 1

emission from radiator = 0.900 kW
emission from pipework = 0.090 kW
(10 per cent of 0.9 kW)

$$\text{total} = 0.990 \text{ kW}$$

$$\text{flow rate} = \frac{0.99}{4.2 \times (80-69)}$$

$$= 0.0214 \text{ kg/s}$$

From the pipe-sizing table (Table 8.5), an 8 mm outside diameter tube will carry 0.0217 kg/s giving a pressure loss per metre run of 902 Pa.

Bed 1
2
1.8 kW

3.000

3.000

2.000

Bed: 2
3
1.7 kW

Manifold

3.000

3.000

3.000

Bed: 3
0.9 kW

4

2.000

Double entry radiator valve

Boiler

Dining room

9

8

7

7

Hall
1.5 kW

Manifold

2 kW
5

3.000

3.000

3.000

4.000

5

6

Index radiator
2.3 kW
6

3.000

Lounge

Note: All pipes assumed uninsulated : hot water cylinder omitted

Fig. 8.3 Micro-pipe system

Pipes number 2

emission from radiator = 1.800 kW
emission from pipework = 0.180 kW
(10 per cent of 1.8 kW)

total = 1.980 kW

flow rate = $\dfrac{1.98}{4.2 \times 11}$

= 0.043 kg/s

From the pipe-sizing Table 8.5, a 10 mm outside diameter tube will carry 0.0433 kg/s giving a pressure loss per metre run of 902 Pa.

Pipes number 3

emission from radiator = 1.700 kW

Table 8.5 Table for the flow of water at 82 °C in small-diameter copper pipes (outside diameter)

Diameter 6 mm	Diameter 8 mm	Diameter 10 mm	Velocity	Pressure drop (Pa per m run of pipe)
(kg/s)	(kg/s)	(kg/s)	(m/s)	
0.005 6	0.013 8	0.027 7		410.0
0.006 3	0.015 3	0.030 6		492.0
0.006 8	0.016 8	0.033 5		574.0
0.007 4	0.018 2	0.036 2	0.6	656.0
0.007.9	0.019 5	0.038 8		738.0
0.008 4	0.020 6	0.041 0		820.0
0.008 8	0.021 7	0.043 3		902.0
0.009 3	0.022 9	0.045 6		984.0
0.009 6	0.024 0	0.047 6		1066.0
0.010 2	0.025 1	0.049 8	0.9	1148.0
0.010 5	0.026 1	0.051 6		1230.0
0.010 9	0.027 1	0.053 5		1312.0
0.011 3	0.027 9	0.055 4		1394.0
0.011 6	0.029 0	0.057 4		1476.0
0.012 1	0.029 8	0.058 7		1558.0
0.012 4	0.030 6	0.060 8		1640.0
0.013 3	0.032 8	0.065 1		1845.0
0.014 2	0.034 9	0.069 0		2050.0
0.015 0	0.036 8	0.072 9	1.2	2255.0

emission from pipework = 0.170 kW
(10 per cent of 1.7 kW)

total = 1.870 kW

flow rate $= \dfrac{1.870}{4.2 \times 11}$

= 0.040 kg/s

From the pipe-sizing Table 8.5, a 10 mm outside diameter tube will carry 0.041 kg/s giving a pressure loss per metre run of 820 Pa.

Pipes number 4

These pipes will have to carry the heat loads required for pipes 1, 2 and 3 plus the heat emission from the pipes number 4.

heat load pipes 1 = 0.990 kW
heat load pipes 2 = 1.980 kW
heat load pipes 3 = 1.870 kW

= 4.840 kW

emission from pipes number 4 = 0.242
(5 per cent of 4.84 kW)

$$\text{total} = 5.082 \text{ kW}$$

$$\text{flow rate} = \frac{5.082}{4.2 \times 11}$$

$$= 0.110 \text{ kg/s}$$

From pipe-sizing Table 8.6, a 22 mm outside diameter tube will carry 0.111 kg/s giving a pressure loss per metre run of pipe of 80 Pa.

Pipes number 5

$$
\begin{aligned}
\text{emission from radiator} &= 2.00 \text{ kW}\\
\text{emission from pipes} &= 0.20 \text{ kW}\\
\text{(10 per cent of 2 kW)} &\\
\hline
\text{total} &= 2.20 \text{ kW}
\end{aligned}
$$

$$\text{flow rate} = \frac{22}{4.2 \times 11}$$

$$= 0.048 \text{ kg/s}$$

From the pipe-sizing Table 8.5, a 10 mm outside diameter tube will carry 0.0498 kg/s giving a pressure loss per metre run of 1148 Pa.

Pipes number 6

$$
\begin{aligned}
\text{emission from radiator} &= 2.300 \text{ kW}\\
\text{emission from pipes} &= 0.230 \text{ kW}\\
\text{(10 per cent of 2.3 kW)} &\\
\hline
\text{total} &= 2.530 \text{ kW}
\end{aligned}
$$

$$\text{flow rate} = \frac{2.530}{4.2 \times 11}$$

$$= 0.054 \text{ kg/s}$$

From the pipe-sizing Table 8.5, a 10 mm outside diameter tube will carry 0.0554 kg/s giving a pressure loss per metre run of 1394 Pa.

Pipes number 7

$$
\begin{aligned}
\text{emission from radiator} &= 1.500 \text{ kW}\\
\text{emission from pipes} &= 0.150 \text{ kW}\\
\text{(10 per cent of 1.5 kW)} &\\
\hline
\text{total} &= 1.650 \text{ kW}
\end{aligned}
$$

$$\text{flow rate} = \frac{1.650}{4.2 \times 11}$$

$$= 0.036 \text{ kg/s}$$

From the pipe-sizing Table 8.5, a 10 mm outside diameter tube will carry 0.0362 kg/s giving a pressure loss per metre run of 656 Pa.

Pipes number 8

These will have to carry the heat loads for pipes 5, 6 and 7 plus the heat emission from pipes number 8.

$$\begin{aligned}\text{heat load pipes 5} &= 2.200 \text{ kW}\\ \text{heat load pipes 6} &= 2.530 \text{ kW}\\ \text{heat load pipes 7} &= 1.650 \text{ kW}\\ \text{total} &= 6.380 \text{ kW}\end{aligned}$$

Pipes number 8 are very much shorter in length than pipes 5, 6 and 7, therefore the estimated heat emission from them will be taken as 5 per cent of the above total.

Heat load to be carried by pipes number 8 is therefore:

6.38 kW + 5 per cent of 6.38

$$\begin{aligned}&= 6.38 + 0.319\\ &= 6.699 \text{ kW}\\ \text{flow rate} &= \frac{6.699}{4.2 \times 11}\\ &= 0.145 \text{ kg/s}\end{aligned}$$

From the pipe-sizing Table 8.6, a 22 mm outside diameter tube will carry 0.152 kg/s giving a pressure loss per metre run of 140 Pa.

Table 8.6 Table for the flow of water at 80 °C in copper pipes (outside diameter). Block table to BS 2871-1-X

Diameter 12 mm	Diameter 15 mm	Diameter 22 mm	Diameter 28 mm	Diameter 35 mm	Velocity	Pressure drop (Pa per m run of pipe)
(kg/s)	(kg/s)	(kg/s)	(kg/s)	(kg/s)	(m/s)	
0.016	0.030	0.089	0.181	0.181	0.3	55.0
0.017	0.032	0.094	0.191	0.344		60.0
0.017	0.033	0.098	0.199	0.360		65.0
0.018	0.035	0.103	0.208	0.376		70.0
0.019	0.036	0.107	0.216	0.390		75.0
0.020	0.037	0.111	0.224	0.405		80.0
0.021	0.040	0.119	0.240	0.433	0.5	90.0
0.022	0.043	0.126	0.254	0.459		100.0
0.025	0.047	1.140	0.282	0.508		120.0
0.027	0.052	0.152	0.308	0.554		140.0
0.030	0.056	0.164	0.332	0.597		160.0
0.032	0.060	0.176	0.354	0.638		180.0
0.034	0.064	0.186	0.376	0.676		200.0
0.038	0.071	0.207	0.416	0.749	1.0	240.0

Pipes number 9

These will have to carry the heat loads for pipes 1, 2, 3, 5, 6, and 7 plus the heat emission from pipes number 4 and 8.

$$\begin{aligned}\text{heat load pipes 1} &= 0.990 \text{ kW}\\ \text{heat load pipes 2} &= 1.980 \text{ kW}\end{aligned}$$

```
            heat load pipes 3 = 1.870 kW
            heat load pipes 4 = 0.242 kW
            heat load pipes 5 = 2.200 kW
            heat load pipes 6 = 2.530 kW
            heat load pipes 7 = 1.650 kW
            heat load pipes 8 = 0.319 kW
                                ─────────
                                11.781 kW
```

Pipes number 9 are again very short and an estimated heat loss from them of 5 per cent of the above total will be adequate.

$$\begin{aligned}
\text{heat load to be carried} \\
\text{by pipes number 9} &= 11.781 + 5 \text{ per cent of } 11.781 \\
&= 11.781 + 0.589 \\
&= 12.370 \text{ kW}
\end{aligned}$$

$$\begin{aligned}
\text{flow rate} &= \frac{12.37}{4.2 \times 11} \\
&= 0.267 \text{ kg/s}
\end{aligned}$$

From the pipe-sizing Table 8.6, a 28 mm outside diameter tube will carry 0.282 kg/s giving a pressure loss per metre run of 120 Pa. A check may be made on the flow rate through pipes number 9 as follows:

$$\begin{aligned}
\text{flow rate through} \\
\text{pipes number 4} &= 0.110 \text{ kg/s} \\
\text{flow rate through} \\
\text{pipes number 8} &= 0.145 \text{ kg/s} \\
&= \underline{0.255 \text{ kg/s}}
\end{aligned}$$

To this flow rate an allowance must be made for the heat emission from pipes number 9.

$$\begin{aligned}
\text{heat loss from pipes number 9} &= 0.589 \text{ kW} \\
\text{(5 per cent of } 11.781 \text{ kW)}
\end{aligned}$$

$$\begin{aligned}
\text{flow rate} &= \frac{0.589}{4.2 \times 11} \\
&= 0.0127 \text{ kg/s}
\end{aligned}$$

$$\begin{aligned}
\text{total flow rate through} \\
\text{pipes number 9} &= 0.255 + 0.0127 \\
&= 0.267 \text{ kg/s}
\end{aligned}$$

This corresponds with the flow rate previously calculated.

Boiler power

The boiler power must be sufficient to supply heat for the radiators, emission from all the pipes, and to heat the water in the storage calorifier.

$$\begin{aligned}
\text{power required for radiators} \\
\text{and pipes} &= 12.370 \text{ kW} \\
\text{power required for hot water} \\
\text{supply} &= \underline{3.000 \text{ kW}} \\
\text{total} &= \underline{15.370 \text{ kW}}
\end{aligned}$$

A 10 per cent boiler margin may be added for pre-heating and to give a reserve of power for severe weather conditions.

Therefore the boiler power would be:

$$15.37 \text{ kW} + 10 \text{ per cent} = 16.907 \text{ kW}$$

A 17 kW boiler would be suitable.

Pump duty

The effective length of the index circuit must first be found by adding an allowance for fittings of 30 per cent on the mains and 10 per cent on the micro-pipe circuit. The index circuit will be the one supplying radiator number 6.

effective length of 10 mm tube

$$14 \text{ m} + 10 \text{ per cent} = 15.4 \text{ m}$$
$$\text{resistance} = 15.4 \times 1.394 \text{ kPa}$$
$$= 21.46 \text{ kPa}$$

effective length of 22 mm and 28 mm tube

$$7 \text{ m} + 30 \text{ per cent} = 9.1 \text{ m}$$
$$\text{resistance} = 9.1 \times 0.140 \text{ kPa}$$
$$= 1.274 \text{ kPa}$$
$$\text{total resistance} = 21.46 + 1.274$$
$$= 22.734 \text{ kPa}$$

A pump providing a pressure of 23 kPa and giving a mass flow rate of 0.3 kg/s would be satisfactory.

Pipe-sizing for gravity circulation

Gravity-circulation systems are not nowadays used for space-heating systems for anything other than a very small building. The system is, however, still used extensively for the primary circulation between the boiler and calorifier in a hot-water supply installation.

Circulating pressure

The pressure required to ensure circulation in gravity systems is termed 'circulation pressure' and is expressed in pascals per metre of circulating height, i.e. the height measured between the centres of the flow and return connections to the boiler and radiator respectively (see Fig. 8.4).

The circulating pressure per metre of height is found from the following formula:

$$\text{circulating pressure} = 9.81 \, (\rho_2 - \rho_1)$$
(Pa per metre height)

where $\rho_2 = $ density of return water/kg/m^3)

$\rho_1 = $ density of flow water (kg/m^3)

Fig. 8.4 Circulation by natural convection

Example 8.4. *Calculate the circulating pressure in pascals when the flow and return water are 82.2 °C and 60 °C respectively and the circulating height is 3m. See Table 8.7 for density of water at various temperatures.*

$$\text{total circulating pressure} = H \times 9.81\,(\rho_2 - \rho_1)$$
$$= 3 \times 9.81\,(983.24 - 970.43)$$
$$= 3 \times 9.81 \times 12.81$$
$$= 376.998 \text{ Pa}$$

This circulating pressure must be sufficient to overcome the pressure due to frictional resistance of the pipe and fittings.

Circulating pressure per metre run

In order to find a suitable diameter of pipe from Table 8.8, it is necessary to find the circulating pressure per metre run of pipe. The run of pipe is the actual length plus the equivalent length due to the resistances of fittings.

157

Table 8.7 Density of water at various temperatures

Temperature (°C)	Density (kg/m³)
43.3	990.93
48.9	988.56
54.4	986.05
60.0	983.24
65.6	980.29
71.0	977.13
76.7	973.88
82.2	970.43
87.8	966.82
93.3	963.07

Table 8.8 Table for the flow of water at 75 °C in black steel pipes. Heavy-grade pipes to BS 1387

Diameter 20 mm		Diameter 25 mm		Diameter 32 mm			Pressure drop (Pa per m run of pipe)
M	EL	M	EL	M	EL	V	
0.020	0.4	0.037	0.7	0.082	1.0		4.0
0.021	0.5	0.039	0.7	0.087	1.0		4.5
0.022	0.5	0.042	0.7	0.093	1.0		5.0
0.023	0.5	0.044	0.7	0.098	1.0		5.5
0.025	0.5	0.046	0.7	0.103	1.1		6.0
0.026	0.5	0.048	0.7	0.107	1.1		6.5
0.027	0.6	0.050	0.7	0.112	1.1		7.0
0.028	0.5	0.052	0.7	0.116	1.1		7.5
0.029	0.5	0.054	0.7	0.120	1.1		8.0
0.030	0.5	0.056	0.7	0.125	1.1		8.5
0.031	0.5	0.058	0.7	0.129	1.1		9.0
0.032	0.5	0.060	0.7	0.133	1.1		9.5
0.033	0.5	0.062	0.7	0.136	1.1	0.15	10.0
0.037	0.5	0.070	0.8	0.154	1.1		12.0
0.042	0.6	0.077	0.8	0.171	1.2		15.0
0.045	0.6	0.084	0.8	0.186	1.2		17.5
0.049	0.6	0.091	0.8	0.200	1.2		20.0
0.052	0.6	0.097	0.8	0.214	1.2		22.5
0.055	0.6	0.103	0.8	0.226	1.2	.030	25.0

V = velocity of flow m/s
M = mass rate of flow kg/s
EL = equivalent length factor

Example 8.5 (see Fig. 8.5). *Find by use of Table 8.8 the diameter of the primary flow and return pipes for a hot-water cylinder holding a mass of 136 kg of water which is to be raised in temperature from 10 °C to 71 °C in 2 hours. The temperatures of the flow and return pipes are to be 71 °C and 60 °C respectively and the circulating height 1.5 m. An allowance of 20 per cent for heat losses on the pipes and a frictional loss of 10 per cent on the actual length of pipes may be used.*

Expansion and feed cistern

Vent pipe

Capacity of calorifier 136 litre

Heating coil

Primary flow pipe

71°C

60°C

1.500

Primary return pipe

1.000

Fig. 8.5 Primary circuit

$$\text{power} = \frac{\text{s.h.c.} \times \text{kg} \times (\text{temperature rise K}) \times 100}{3600 \times 2 \times \text{efficiency}}$$

$$= \frac{4.2 \times 136 \times (71-10) \times 100}{3600 \times 2 \times 80}$$

$$= 6.049 \text{ kW}$$

159

$$\text{flow rate} = \frac{6.049}{4.2 \times 11}$$

$$= 0.131 \text{ kg/s}$$

$$\text{total circulating pressure} = H \times 9.81\,(\rho_2 - \rho_1)$$

$$= 1.5 \times 9.81 \times (983.24 - 977.13)$$

$$= 1.5 \times 9.81 \times 6.11$$

$$= 90 \text{ Pa}$$

$$\text{effective length of pipe} = 5 + 10 \text{ per cent}$$

$$= 5.5 \text{ m}$$

$$\text{circulating pressure available per metre run} = \frac{\text{total circulating pressure}}{\text{effective length}}$$

$$= \frac{90}{5.5}$$

$$= 16.36 \text{ Pa}$$

From the pipe-sizing Table 8.8, a 32 mm diameter steel pipe will carry 0.133 kg/s giving a pressure loss of 9.5 Pa per metre run. Since the flow rate for this pipe is greater than required and the frictional loss is less, the pipe will be satisfactory.

Sizing of secondary circuits

The secondary flow pipe will have to deliver a certain rate of flow to the various sanitary fittings, while the secondary return will have to carry sufficient water to compensate for the heat losses from the pipes and any heat emitted from towel rails, etc. The secondary return may therefore be determined as previously described for the pipe-sizing for heating.

The sizing of the secondary flow pipe can be carried out from charts or tables or by the Box formula described in Chapter 3, using the discharge rates from the sanitary fittings as listed in Table 8.9, and allowing a diversity factor.

Table 8.9   Recommended minimum rate of flow at various appliances

Type of appliance	Rate of flow (litre/s)
WC flushing cistern	0.12
Wash basin	0.15
Wash basin with spray taps	0.04
Bath (private)	0.30
Bath (public)	0.60
Shower (with nozzle)	0.12
Sink with 13 mm taps	0.20
Sink with 19 mm taps	0.30
Sink with 25 mm taps	0.60

Index circuit (see Fig. 8.6)

In a gravity-heating system the index circuit may be defined as the circuit in which the ratio of the circuit length to the circuit circulating height is the minimum.

If the temperature drop across each circuit in Fig. 8.6 is the same, and since circuit 1 has twice the circulating height of circuit 2, it will have double the circulating pressure of circuit 2. Since circuit 2 has a greater length than circuit 1, it follows that circuit 2 is the index circuit.

In order to find the diameters of pipes for the system, the circulating pressure per metre of travel will be based on circuit 2. The method of pipe-sizing will then be carried out as described for the primary flow and return pipes in a hot-water supply system.

Fig. 8.6 Index circuit

Questions

1. Figure 8.7 shows a single-pipe circuit, assuming each radiator has a heat emission of 5 kW and the temperature drop across the system is 10 °C. Find by use of Table 8.10 the approximate diameter of a copper main pipe. A velocity of flow of water is to be 0.5 m/s. Determine the pump duty for the circuit allowing an emission from the pipes 20 per cent of the total from the radiators.

161

Fig. 8.7

Table 8.10   Table for the flow of water at 75 °C in copper tubes BS 659
Light Gauge

Diameter 51 mm		Diameter 63 mm		Diameter 76 mm			Pressure drop (Pa per m run of pipe)
M	EL	M	EL	M	EL	· V	
0.428	2.2	0.774	2.9	1.26	3.8	0.30	10.0
0.486	2.2	0.878	3.0	1.42	3.8		12.5
0.538	2.3	0.972	3.1	1.58	3.9		15.0
0.587	2.3	1.06	3.1	1.72	4.0		17.5
0.633	2.4	1.14	3.2	1.85	4.1		20.0
0.676	2.4	1.22	3.2	1.98	4.1		22.5
0.714	2.4	1.29	3.3	2.10	4.2		25.0
0.757	2.5	1.37	3.3	2.21	4.2		27.5
0.795	2.5	1.43	3.3	2.32	4.3		30.0
0.838	2.5	1.50	3.4	2.42	4.3		32.5
0.867	2.5	1.56	3.4	2.53	4.3		35.0
0.907	2.5	1.62	3.4	2.63	4.4	0.50	37.5

V = velocity of flow m/s
M = mass rate of flow kg/s
EL = equivalent length factor

2. Figure 8.8 shows a two-pipe circuit assuming each radiator has a heat emission of 8 kW and the temperature drop across the system is 10 °C. Find by use of Table 8.11 the approximate diameters of the steel pipes. Determine the pump duty for the circuit allowing an emission from each section of the pipework 5 per cent of the total from the radiators.

Answers: pipe number 1, 20 mm diameter; pipe number 2, 25 mm diameter; pump duty – pressure 5 kPa; discharge 0.3 kg/s.

Fig. 8.8

Table 8.11   Table for the flow of water at 75 °C in black steel pipes. Heavy-grade pipes to BS 1387

Diameter 15 mm		Diameter 20 mm		Diameter 25 mm			Pressure drop (Pa per m run of pipe)
M	EL	M	EL	M	EL	V	
0.011	0.3	0.026	0.5	0.048	0.7	0.15	6.5
0.011	0.3	0.027	0.5	0.050	0.7	0.15	7.0
0.012	0.3	0.028	0.5	0.052	0.7	0.15	7.5
0.012	0.3	0.029	0.5	0.054	0.7	0.15	8.0
0.013	0.3	0.030	0.5	0.056	0.7	0.15	8.5
0.013	0.3	0.031	0.5	0.058	0.7	0.15	9.0
0.014	0.3	0.032	0.5	0.060	0.7	0.15	9.5
0.014	0.3	0.033	0.5	0.062	0.7	0.15	10.0
0.016	0.3	0.037	0.5	0.070	0.8	0.15	12.5
0.018	0.4	0.042	0.6	0.077	0.8	0.15	15.0
0.019	0.4	0.045	0.6	0.084	0.8	0.15	17.5
0.021	0.4	0.049	0.6	0.091	0.8	0.15	20.0

V = velocity of flow m/s
M = mass rate of flow kg/s
EL = equivalent length factor

3. Define the following terms: (a) index circuit; (b) index radiator; (c) circulating pressure; (d) circulating height.

4. Figure 8.9 shows a part of a micro-pipe system which represents the index circuit. If an 'open' type of system is to be used with flow and return water temperatures of 80 °C and 69 °C respectively, find the diameters of the micro-pipes and the total pressure drop in pascals through them.

Answers: 10 mm O.D. copper tube; pressure drop 24 805 Pa or 24.8 kPa.

Fig. 8.9

5. Calculate the total circulating pressure for a gravity heating system when the circulating height is 4 m and the flow and return waters are 82.2 °C and 60 °C respectively.

Answer: 502.3 Pa

6. Find by the use of Table 8.8 suitable diameter flow and return pipes for a 136 litre hot-water cylinder. Use the following factors:

(a) heat recovery period = 3 hours
(b) circulating height = 1.2 m
(c) actual length of pipes = 4.4 m
(d) temperature flow of water = 71 °C
(e) temperature of return water = 60 °C
(f) temperature rise of water in the cylinder = 65 °C
(g) heat loss in pipes = 20 per cent of power required to heat water
(h) frictional loss on pipes = 10 per cent of actual length

Answer: 32 mm diameter steel pipe

Chapter 9

Ventilating ducts and fans

Ventilating ducts

The sizing of ducts to convey air from fans in a plant room to the various spaces inside a building relies upon empirical data for the evaluation of frictional losses. It cannot therefore be assumed that the calculated volumes of air will correspond with those actually delivered to the various discharge points, and for this reason dampers are incorporated inside the duct system. These dampers perform the same function as regulating valves in hot-water heating systems.

Flow of air in ducts

The Bernoulli theorem may be applied to air flow in ducts provided that an allowance is made for friction and separation. The theorem states that the total energy of each particle of a fluid is the same, provided that no energy enters or leaves the system at any point. If there is a loss of one type of energy, there must be a corresponding gain in another, or vice versa.

Figure 9.1 shows the total energies of air flowing inside a section of ductwork. The total energy possessed by the air is the sum of the potential, pressure and kinetic energies and this can be expressed as follows:

$$\text{total energy } H = z + \frac{P}{\rho g} + \frac{V^2}{2g} = \text{constant}$$

where z = potential energy due to the position of the particles or height above the datum (m)

$$\frac{P}{\rho g} = \text{pressure energy due to the depth of the particles below the fluid (m)}$$

$$\frac{V^2}{2g} = \text{kinetic energy due to the movement of particles of the fluid (m)}$$

Fig. 9.1 Total energy of a moving fluid

Because the mass of air involved and its density are small, the value of the potential energy will also be small and for practical purposes is neglected. This reduces the equation to:

$$H = \frac{P}{\rho g} + \frac{V^2}{2g} = \text{constant}$$

Pressure distribution

In a closed system the total energy at two points must be equal, provided that there are no losses or gains.

$$\frac{P_1}{\rho g} + \frac{V_1^2}{2g} = \frac{P_2}{\rho g} + \frac{V_2^2}{2g}$$

If ideal conditions existed, i.e. no pressure loss due to friction, the distribution of pressure would be as shown in Fig. 9.2.

The loss of pressure energy due to friction between two points in a duct may be described in diagrammatic form (see Fig. 9.3).

The pressure energy $P/\rho g$ is referred to as the static head, or pressure of the system, and may be defined as the pressure acting equally in all directions. It is the pressure that tends either to burst the duct outwards when it is positive, or to collapse the duct inwards if it is negative.

The velocity energy $V^2/2g$ is referred to as the velocity head, or pressure of the system, and may be defined as the energy required either to accelerate the air from rest to a certain velocity, or to bring the air from a certain velocity to rest.

Pressure distribution
TP = Total pressure
SP = Static pressure
VP = Velocity pressure

Fig. 9.2 Total energy at two points in a ventilating duct for ideal conditions

TP = Total pressure   SP = Static pressure   VP = Velocity pressure

Fig. 9.3 Total energy at two points in a ventilating actual duct for conditions
(i.e. pressure loss due to friction)

Measurement of pressure (see Fig. 9.4)

In order to measure the pressures inside a duct, a Pitot tube is used. One mano-meter will show the total pressure while the other will show the static pressure.

Fig. 9.4 Pitot tube

The velocity pressure may be found by deducting the static pressure from the total pressure, i.e.

$$\text{velocity pressure or head} = \text{total pressure or head} - \text{static pressure or head}$$

Because the manometer-tube readings are very small, in order to obtain a higher degree of accuracy, an inclined manometer may be used with the Pitot tube.

The velocity pressure is always positive and since the total pressure is the sum of the static and velocity pressure, the total pressure may therefore be either positive or negative.

Velocity and velocity head (VH)

$$\text{VH} = \frac{V^2}{2g}$$

since the measurement is expressed in water gauge, and air is being considered:

$$\text{VH} = \frac{V^2}{2g} \times \frac{\rho_a}{\rho_w}$$

where VH = velocity head (m)
V = velocity of air flow (m/s)
ρ_a = density of air (1.2 kg/m^3)
ρ_w = density of water (998 kg/m^3)

$$\therefore \text{VH} = \frac{V^2}{2 \times 9.81} \times \frac{1.2}{998}$$

and $= V^2 \times 0.000\,061\,2$ (m) of water

In terms of mm of water,

$$VH = V^2 \times 0.0612$$

and

$$V = \frac{VH}{0.0612}$$

Velocity pressure (VP)

Manufacturers of fans often state the pressure developed by a fan in pascals instead of in millimetres water gauge. The pressure in pascals may be found as follows:

$$\text{pressure (Pa)} = \text{head of water (m)} \times 1000 \times 9.81$$
$$\therefore\ VP = V^2 \times 0.000\,061\,2 \times 1000 \times 9.81$$
$$= V^2 \times 0.6 \text{ Pa (approx.)}$$

This is the form given in the Chartered Institution of Building Services guide book and a table listing velocity pressure in pascals against velocity in m/s is produced.

Example 9.1. *Calculate the velocity pressure in pascals in a ventilating duct when the velocity of air is found to be 5 m/s.*

$$VP = 5^2 \times 0.6$$
$$= 15 \text{ Pa}$$

Example 9.2. *If the velocity pressure of air flowing in a duct is found to be 10 pascals, calculate the velocity of the air.*

$$VP = V^2 \times 0.6$$

$$V = \sqrt{\frac{VP}{0.6}}$$

$$V = 4 \text{ m/s (approx.)}$$

Example 9.3. *Calculate the velocity head in mm water gauge when the velocity of flow of air in a duct is 4 m/s.*

$$VH = 4^2 \times 0.0612$$
$$= 0.979 \text{ mm (approx.)}$$

Example 9.4. *Calculate the theoretical velocity of the flow of air in a duct when the velocity head is found to be 4 mm water gauge.*

$$V = \sqrt{\frac{4}{0.0612}}$$

$$= 8 \text{ m/s (approx.)}$$

Note: In practice, the velocity of air flow in a duct is often found by means of a velometer or anemometer.

Volume of air flow

The following formula may be used:

$$Q = VA$$

where Q = volume of air flow (m³/s)
V = velocity of air flow (m/s)
A = cross-sectional area of duct (m²)

Example 9.5. *The flow rate in a circular duct is to be determined by means of a Pitot tube. Determine the volume of flow in m³/s from the following data:*

diameter of duct = 600 mm
total head (TH) = 30 mm water gauge
static head (SH) = 18 mm water gauge
standard air density = 1.20 kg/m³ at 20°C

$$VH = TH - SH$$
$$= 30 - 18$$
$$= 12 \text{ mm water gauge}$$

$$V = \sqrt{\frac{12}{0.0612}}$$

$$V = 14 \text{ m/s}$$
$$Q = VA$$
$$Q = V \times 0.7854 \, D^2$$
$$Q = 14 \times 0.7854 \times 0.600 \times 0.600$$
$$Q = 3.96 \text{ m}^3/s \text{ (approx.)}$$

Example 9.6. *A room measuring 20 m × 10 m × 3 m requires ventilating by means of a fan and ductwork to provide six air changes in the room. If the average velocity of air flow in the duct is to be 2 m/s, calculate the diameter of the main circular duct required for the room.*

$$Q = \frac{20 \times 10 \times 3 \times 6}{3600}$$

$$Q = 1 \text{ m}^3/s$$
$$Q = V \times 0.7854 \, D^2$$

$$D = \sqrt{\frac{Q}{V \times 0.7854}}$$

$$D = \sqrt{\frac{1}{2 \times 0.7854}}$$

$$D = 0.798 \text{ m or } 798 \text{ mm}$$

The D'Arcy formula

It was shown in Chapter 3 that the loss of head due to friction for water flowing through a pipe was given in the following expression:

$$b = \frac{4fL\,V^2}{2\,g\,D}$$

The formula may be used for small circular ventilating ducts in the following terms:

$$b = \frac{4fL\,V^2}{2\,g\,D} \times \frac{\rho_a}{\rho_w}$$

where b = loss of static head due to friction (m)
f = coefficient of friction
L = length of duct (m)
V = velocity of flow (m/s)
D = diameter of duct (m)
ρ_a = density of air (1.2 kg/m^3)
ρ_w = density of water (998 kg/m^3)

$$\therefore\; b = \frac{4fL\,V^2}{2\,g\,D} \times 0.0012$$

The coefficient of friction depends upon the condition of the internal surface of the duct, and Reynolds number. For practical purposes, a value of between 0.005 and 0.007 may be used. The type of flow as well as the type of fluid affects the value, so that the D'Arcy formula acts as a guide only to the rate of frictional loss.

Example 9.7. *Calculate the static head lost due to friction in a 150 mm diameter ventilating duct 12 m long when the average velocity of air flow through the duct is 8 m/s.*

$$b = \frac{4 \times 0.007 \times 12 \times 8^2 \times 0.0012}{2 \times 9.81 \times 0.150}$$

h = 0.008 77 m or 877 mm

Rectangular ducts

In order to find the head lost due to friction, the D'Arcy formula may be modified as follows:

$$b = \frac{2\,(a+b)fL\,V^2}{2\,g\,a\,b} \times \frac{\rho_a}{\rho_w}$$

$$h = \frac{2\,(a+b)fL\,V^2}{2\,g\,a\,b} \times 0.0012$$

where a and b are the lengths of the sides of the rectangular duct in metres.

Example 9.8. *Calculate the static head lost due to friction in a rectangular ventilating duct having sides measuring 450 mm by 400 mm. The length of the duct is 30 m and the average velocity of air flowing through it 5 m/s.*

$$b = \frac{2\,(0.45+0.40) \times 0.007 \times 30 \times 5^2 \times 0.0012}{2 \times 9.81 \times 0.45 \times 0.40}$$

h = 0.003 m or 3 mm water gauge

171

Note: It is sometimes necessary to find the diameter of a circular duct having the same cross-sectional area as a rectangular duct, or vice versa. The diameter of the circular duct may be found from one of the following formulae, depending upon the problem in hand. Tables are also available for this purpose.

For equal velocity of flow:

$$d = \frac{2\,a\,b}{(a+b)}$$

For equal volume of flow:

$$d = 1.265 \times \left[\frac{(a\,b)^3}{a+b} \right]^{0.2}$$

where *a* and *b* are the lengths of the sides of the rectangular duct.

Duct sizing by use of chart (see Fig. 9.5)

In practical duct sizing, a chart is used which is derived from a more sophisticated formula than the D'Arcy, i.e. Colebrook-White equation.

Example 9.9. *Find by the use of the chart (see Fig. 9.5) the diameter of a circular duct 12 m long that will give a flow rate of 1 m³/s when the velocity of flow is 8 m/s. Find the static head lost due to friction in mm water gauge.*

Figure 9.6 shows how the chart should be used and the following values are found:

1. diameter of duct   = 400 mm
2. frictional loss per metre run = 0.24 mm
3. total frictional loss   = 0.24 × 12
                                = 2.88 mm

This may be used as a check against the D'Arcy formula previously explained:

$$h = \frac{4 \times 0.007 \times 12 \times 8^2 \times 0.0012}{2 \times 9.81 \times 0.400}$$

h = 0.003 287 m or 3.2 mm

The D'Arcy formula gives 0.32 mm higher frictional loss, which is negligible in this example.

Loss of head or pressure due to fittings

The resistance to the flow of air in ducts due to fittings such as bends, branches, dampers, etc., may be expressed in head of water and this value may then be converted into pressure in pascals, as described previously. The fundamental expression for the velocity head for the flow of water is given in the following expression:

$$h = \frac{V^2}{2g}$$

For the flow of air in ducts this becomes

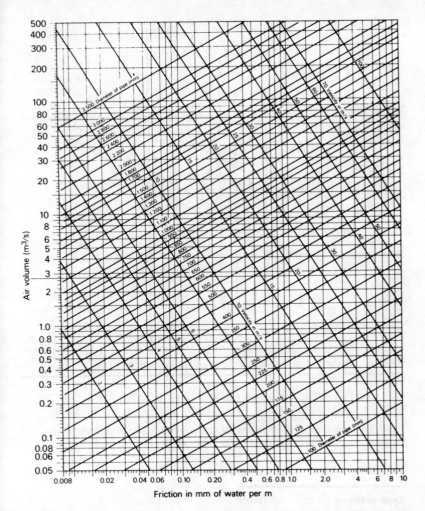

Fig. 9.5 Duct-sizing chart

$$b = \frac{V^2}{2g} \times \frac{\rho_a}{\rho_w}$$

and expressing the resistance of fittings in terms of velocity head, the following expression results:

$$b = k \quad \frac{V^2}{2g} \quad \frac{\rho_a}{\rho_w}$$

where b = velocity head loss due to fittings (m)
k = velocity head factor determined experimentally
V = velocity of air flow (m/s)

173

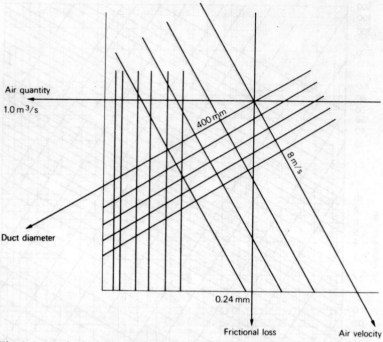

Air quantity

1.0 m³/s

400 mm

8 m/s

Duct diameter

0.24 mm

Frictional loss

Air velocity

Fig. 9.6 Method of using duct-sizing chart for example 9.8

$$\rho_a = \text{density of air } (1.2 \text{ kg/m}^3 \text{ at } 20\,^\circ C)$$
$$\rho_w = \text{density of water } (998 \text{ kg/m}^3)$$

Table 9.1 gives some general values for velocity-head factors for illustration purposes only. A full list is given in the CIBS guide book.

Table 9.1 Velocity-head loss factors (k)

Type of fitting	k factor
90° Rounded elbow	0.65
90° Square elbow	1.25
90° Bend ($R = 2\,D$)	0.10
Branch	0.6–1.3
Abrupt enlargement	0.30
Gradual enlargement	0.15
Abrupt reducing	0.30
Gradual reducing	0.04
Open damper	0.30
Wire mesh	0.40
Diffuser	0.60
Outlet	1.00

Example 9.10. *Determine the total loss of head in mm water gauge in a 300 mm diameter duct 15 m long, having three 90° bends when the velocity of air flow is 6 m/s.*

$$\text{velocity-head loss due to bends} = 3 \times 0.1 \left[\frac{6^2}{2 \times 9.81} \right] \times 0.0012$$

$$= 0.000\ 66 \text{ m}$$

$$= 0.66 \text{ mm}$$

Static-head loss due to straight duct, using the duct-sizing chart (Fig. 9.5) for 300 mm diameter and 6 m/s.

$$\text{for 1 m} = 0.19 \text{ mm}$$

$$\text{for 15 m} = 0.19 \times 15$$

$$= 2.85 \text{ mm}$$

$$\text{total head loss} = 0.66 + 2.85$$

$$= 3.51 \text{ mm}$$

Sizing of system of ductwork

Several methods may be used to determine the sizes of ventilating ducts to deliver or extract air from various spaces inside a building. In theory, duct sizing is very complex, but the following approximate methods are used in practice:

Equal-velocity method: The air velocity in the ductwork is decided by the designer and duct cross-sectional areas obtained from the volumes of flows required using the continuity equation. This method is generally used for simple systems without branches and for pneumatic conveying.

Velocity-reduction method: The air velocity in the first section of the main duct is chosen and the velocity reduced at each branch. Duct sizes are then found using the continuity equation.

Static-regain method: The size of a duct between two adjacent branches is chosen so that the pressure drop over the section is equal to the static regain at the previous branch. Therefore the static pressure at every branch is the same.

Constant pressure drop per unit length: The air velocity in the first section of the main duct is chosen and the size of the duct in this section found. Using the duct-sizing chart, the pressure drop per unit length of duct is obtained for this section and this value is then used for all the remaining sections of ductwork. This is a very convenient method of duct sizing and will be the one used in the following example.

For low-velocity systems a total pressure drop of 1 Pa may be used.

Example 9.11. *Figure 9.7 shows an extract system of ductwork. Determine by use of the duct-sizing chart (Fig. 9.5) the diameters of the ducts A, B and C assuming that the average velocity of air flow through duct A is to be 5 m/s.*

Fig. 9.7 System of ductwork

Duct A

$$\text{volume of flow} = 0.5 + 0.6 + 0.3$$
$$= 1.4 \text{ m}^3/\text{s}$$

From the duct-sizing chart a flow rate of 1.4 m³/s and an air velocity of 5 m/s gives a head loss per metre run of 0.059 mm water gauge and a 600 mm diameter duct.

Duct B

Using the same head loss per metre run as for duct A,

$$\text{flow rate} = 0.6 + 0.3$$
$$= 0.9 \text{ m}^3/\text{s}$$

From the duct-sizing chart a 500 mm diameter duct giving an air velocity of approximately 4.4 m/s is found to be suitable.

Duct C

Using the same head loss per metre run as for duct A,

$$\text{flow rate} = 0.3 \text{ m}^3/\text{s}$$

From the duct-sizing chart a 350 mm diameter duct giving an air velocity of approximately 3.4 m/s is found to be suitable.

Total head loss

The values of resistances through the various fittings and the lengths of duct-work can now be calculated basing the velocity head on the velocity through each section of ductwork. The total loss of pressure through the duct will be required when ordering the fan.

Duct A

$$\text{loss of head through straight duct} = 8 \times 0.059$$
$$= 0.422 \text{ mm}$$
$$= 0.000\ 422 \text{ m}$$

To this must be added the loss of head through the outlet, which from Table 9.1 has a *k* factor of 1.0.

$$\text{loss through outlet} = 1 \left[\frac{5^2}{2 \times 9.81} \right] \times 0.0012$$
$$= 0.0015 \text{ m}$$
$$\text{total loss} = 0.000\ 422 + 0.0015$$
$$= 0.001\ 922 \text{ m}$$
$$= 0.002 \text{ m (approx.)}$$

176

Duct B

loss of head through straight duct = 10×0.059

$$= 0.59 \text{ mm}$$

$$= 0.000\ 59 \text{ m}$$

To this must be added resistances of two branches and two inlets.

$$\text{branches (average } k, 0.9) = 2 \times 0.9 \left[\frac{4.4^2}{2 \times 9.81} \right] \times 0.0012$$

$$= 2 \times 0.9 \times 0.987 \times 0.0012$$

$$= 0.002\ 13 \text{ m}$$

$$\text{inlets (wire mesh } k, 0.4) = 2 \times 0.4 \left[\frac{4.4^2}{2 \times 9.81} \right] \times 0.0012$$

$$= 2 \times 0.4 \times 0.987 \times 0.0012$$

$$= 0.000\ 947 \text{ m}$$

$$\text{total loss} = 0.000\ 59 + 0.002\ 13 + 0.000\ 947$$

$$= 0.003\ 667 \text{ m}$$

$$= 0.004 \text{ m (approx.)}$$

Duct C

loss of head through straight duct = 10×0.059

$$= 0.59 \text{ mm}$$

$$= 0.000\ 59 \text{ m}$$

To this must be added one bend and one inlet.

$$\text{bend } (k, 0.1) = 0.1 \left[\frac{3.4^2}{2 \times 9.81} \right] \times 0.0012$$

$$= 0.1 \times 0.589 \times 0.0012$$

$$= 0.000\ 070\ 6 \text{ m}$$

$$\text{inlet } (k, 0.4) = 0.4 \left[\frac{3.4^2}{2 \times 9.81} \right] \times 0.0012$$

$$= 0.4 \times 0.589 \times 0.0012$$

$$= 0.000\ 287 \text{ m}$$

$$\text{total loss} = 0.000\ 59 + 0.000\ 07 + 0.000\ 287$$

$$= 0.000\ 947 \text{ m}$$

$$\text{loss through duct system} = 0.002 + 0.004 + 0.0095$$

$$= 0.0155 \text{ m}$$

Converting this loss of head to pressure (Pa)

$$\text{pressure} = 0.0155 \times 1000 \times 9.81$$

$$= 152 \text{ Pa}$$

Fan duty

A fan that will discharge 1.4 m^3 /s and develop a total negative pressure of

152 Pa would be suitable. Dampers placed in the inlet branches will enable the system to be balanced.

Fans

There are three main types of fans used for ventilation systems:

Propeller fan: This type is usually used at free opening in walls or windows and for other types of low pressure applications. It is not usually employed for ducted ventilation systems.

Axial-flow fan: This type of fan is designed for mounting inside a duct system and is suitable for moving air in complete systems of ductwork.

Centrifugal fan: The inlet of the fan is at $90°$ to the outlet and like the axial-flow fan is suitable for moving air in complete systems of ductwork.

Fan laws

The performance of a fan incorporated in a system of ventilation is governed by the following laws, providing the air density remains constant:

1. The discharge varies directly with the angular velocity of the impeller.
2. The pressure developed varies directly as the square of the angular velocity of the impeller.
3. The power absorbed varies as the cube of the angular velocity of the impeller.

The laws may be expressed as follows:

1. $$\frac{Q_2}{Q_1} = \frac{N_2}{N_1}$$

2. $$\frac{P_2}{P_1} = \frac{(N_2)^2}{(N_1)^2}$$

3. $$\frac{power_2}{power_1} = \frac{(N_2)^3}{(N_1)^3}$$

where Q = volume of flow in m^3/s

N = revolutions of impeller per minute

P = pressure in pascals

power = power in watts or kilowatts

Example 9.12. *A fan absorbs 2.3 kW of power and discharges 2.5 m^3/s when the impeller angular velocity is 1000 revolutions per minute. If the impeller angular velocity is increased to 1200 revolutions per minute, calculate the discharge in m^3/s and the power absorbed for this new condition.*

$$\frac{Q_2}{Q_1} = \frac{N_2}{N_1}$$

$$\therefore Q_2 = \frac{Q_1 \times N_2}{N_1}$$

178

$$Q_2 = \frac{2.5 \times 1200}{1000}$$

$$= 3 \text{ m}^3/\text{s}$$

$$\frac{power_2}{power_1} = \frac{(N_2)^3}{(N_1)^3}$$

$$\therefore power_2 = \frac{W_1 \times (N_2)^3}{(N_1)^3}$$

$$= \frac{2.3 \times 1200^3}{1000^3}$$

$$= 3.9744 \text{ kW}$$

It will be observed from example 9.12 that for a 20 per cent increase in angular velocity there is a 72.8 per cent increase in power absorbed. The common method used to lower the air flow through a main duct is by closing a damper, and this can lead to a significant loss of power. An alternative method of lowering the air flow is by lowering the angular velocity of the impeller. This can be achieved by varying the frequency of the current input to the motor driving the fan.

Example 9.13. *A fan develops a static pressure of 200 Pa when the angular velocity of the impeller is 900 revolutions per minute. If the angular velocity of the impeller is increased to 1000 revolutions per minute, calculate the static pressure developed by the fan for the new condition.*

$$\frac{P_2}{P_1} = \frac{(N_2)^2}{(N_1)^2}$$

$$\therefore P_2 = \frac{P_1 \times (N_2)^2}{(N_1)^2}$$

$$= \frac{200 \times 1000^2}{900^2}$$

$$= 246.9$$
$$= 247 \text{ Pa (approx.)}$$

Efficiency of fans

The efficiency of a fan can be found by use of the following expressions:

$$\frac{\text{efficiency}}{\text{(per cent)}} = \frac{\text{fan total pressure} \times \text{volume of flow}}{\text{power absorbed (W)}} \times \frac{100}{1}$$

Example 9.14. *A fan absorbs 1.5 kW when discharging 2 m^3/s of air and operating at a total pressure of 500 Pa. Calculate the percentage efficiency of the fan.*

$$\text{efficiency} = \frac{500 \times 2}{1500} \times \frac{100}{1}$$

$$= 66.667 \text{ per cent}$$

Change of air density

If the air density is changed, the following laws apply:

1. The volume of flow remains constant.
2. The pressure developed varies directly with the change in density.
3. The power absorbed varies directly with the change in density.

The laws may be expressed as follows:

1. $\quad Q_1 = Q_2$

2. $\quad \dfrac{P_2}{P_1} = \dfrac{\rho_2}{\rho_1}$

3. $\quad \dfrac{power_2}{power_1} = \dfrac{\rho_2}{\rho_1}$

where Q = volume of flow in m^3/s
$\quad\quad\;\, P$ = pressure in pascals, which may be static, velocity or total
$\quad\quad\;\, \rho$ = density of air in kg/m^3
$\quad$ power = power in watts or kilowatts

Example 9.15. *A fan develops a total pressure of 400 Pa when discharging air at a temperature of 20 °C. If the temperature of the air is lowered to 16 °C, calculate the pressure to be developed by the fan in order to discharge the same volume of air.*

density of air at 20 °C = 1.2 kg/m^3
density of air at 16 °C = 1.22 kg/m^3

$$\frac{P_2}{P_1} = \frac{\rho_2}{\rho_1}$$

$$\therefore\ P_2 = \frac{P_1 \times \rho_2}{\rho_1}$$

$$= \frac{400 \times 1.22}{1.2}$$

$$= 406.67\ \text{Pa}$$

Selection of a fan

In order to select a fan for a given duty, reference should be made to fan-performance graphs supplied by the manufacturer.

Figure 9.8 shows a typical performance graph for two fans and it will be seen that the following factors may be obtained from them:

1. pressure in pascals;
2. volume flow rate in m^3/s;
3. input in watts.

Before using the graphs it is essential to draw a system characteristic curve and the intersection of this curve with the fan-performance curve will give the actual amount of discharge and the pressure developed by the fans.

Note: Manufacturers of fans will supply graphs giving pressure, volume of air flow, power and efficiency

Fig. 9.8 Performance graph for fans

System characteristic curve

For a certain rate of flow of air through a system of ductwork a certain static pressure must be developed.

For any system of ductwork therefore, a characteristic curve may be drawn plotting the rate of flow against the static pressure to be developed in order to overcome the resistances.

Example 9.16. *A ventilating system requires a fan to discharge 6 m³/s against a calculated resistance of 300 Pa pressure. Select from the graph in Fig. 9.8 either fan A or fan B.*

For a given system of ductwork the pressure loss may be found from the following expression:

pressure loss = pressure-loss coefficient × volume of flow²

$$P = k \, Q^2$$

$$\therefore k = \frac{P}{Q^2}$$

$$= \frac{300}{6^2}$$

$$= 8.333$$

In order to draw the system characteristic curve it is necessary to find the values of pressure corresponding to nominated values of flow. Using the flow rates between 0.5 and 6 m³/s :

1. $Q = 8.33 \times 0.5^2 = 2.08$ Pa
2. $Q = 8.33 \times 1.0^2 = 8.33$ Pa
3. $Q = 8.33 \times 1.5^2 = 18.74$ Pa
4. $Q = 8.33 \times 2.0^2 = 33.32$ Pa
5. $Q = 8.33 \times 2.5^2 = 52.06$ Pa
6. $Q = 8.33 \times 3.0^2 = 74.97$ Pa
7. $Q = 8.33 \times 3.5^2 = 102.04$ Pa
8. $Q = 8.33 \times 4.0^2 = 133.28$ Pa
9. $Q = 8.33 \times 4.5^2 = 168.68$ Pa
10. $Q = 8.33 \times 5.0^2 = 208.25$ Pa
11. $Q = 8.33 \times 5.5^2 = 251.98$ Pa
12. $Q = 8.33 \times 6.0^2 = 299.88$ Pa

By reference to Fig. 9.8, fan A will discharge 6.5 m³/s and develop a pressure of 320 Pa for the conditions given. Fan B would discharge 5.9 m³/s and develop a pressure of 272 Pa for the same conditions.

The designer has a choice between fan A and fan B, and fan A would probably be chosen. If it is required to save electrical power, however, fan B might be selected.

Limiting velocities in ducts

In order to reduce noise and power consumption, the velocity of air flowing through a duct must be kept within reasonable limits. Table 9.2 gives values of limiting velocities in ducts.

Table 9.2 Limiting velocities in ducts

Application	Velocity (m/s)	
	Commercial buildings	Industrial buildings
Outside air intake	5.0	7.5
Discharge to atmosphere	5.0	7.5
Main supply or extract duct	7.5	13.0
Terminal branch duct	3.5	5.0

Questions

1. Make a statement of Bernoulli's theorem as it applies to air flow in ductwork.

2. Define velocity, static and total pressure and describe how the Pitot tube may be used to measure the velocity, static and total pressures of air flowing through a duct.

3. Calculate the velocity pressure in pascals in a ventilating duct when the velocity of air is found to be 7 m/s.

Answer: 29.4 Pa

4. Calculate the velocity of air flowing in a duct when the velocity pressure is 25 Pa.

Answer: 6.455 m/s

5. Calculate the flow rate in m^3/s through a 350 mm diameter ventilating duct when the total and static heads are 30 mm water gauge and 25 mm water gauge respectively. Assume standard air density of 1.20 kg/m^3.

Answer: 0.87 m^3/s (approx.)

6. A room measuring 15 m × 8 m × 3 m requires ventilating by means of a fan and ductwork to provide three air changes per hour. If the average velocity of air flow in the duct is to be 4 m/s, calculate the diameter of the main circular duct required for the room.

Answer: 309 mm

7. Calculate the static head lost due to friction in a 300 mm diameter ventilating duct 15 m long when the average velocity of air flow through the duct is 5 m/s. Use the following values:

    (*a*) density of water   = 998 kg/m^3
    (*b*) density of air   = 1.2 kg/m^3
    (*c*) coefficient of friction = 0.005

Answer: 1.53 mm water gauge

8. Calculate the static head lost due to friction in a rectangular ventilating duct having sides 350 mm by 300 mm. The length of the duct is 12 m and the average velocity of air flowing through it 3 m/s. Use the values given in Question 7.

Answer: 0.8 mm water gauge

9. Find by use of the duct-sizing chart (Fig. 9.5) the diameter of a circular duct 50 m long that will give a flow rate of 9 m^3/s when the velocity of flow is 5 m/s. Find the total static head lost due to friction in mm water gauge.

Answers: diameter 1.4 m; total static head lost 0.9 mm water gauge

10. Determine the total loss of head in mm water gauge in a 500 mm diameter duct 30 m long, having four 90° bends, when the velocity of air flow is 5 m/s.

Answer: 2.8 mm water gauge (approx.)

11. Figure 9.9 shows an inlet system of ductwork. Determine by use of the duct-sizing chart (Fig. 9.5) the diameters of the ducts A and B. It may be assumed that the average velocity of air flow through duct A is to be 6 m/s.

Answers: duct A 500 mm diameter; duct B 450 mm diameter

Chapter 10

Cold- and hot-water storage, expansion of materials, Boyle's and Charles' laws

Cold-water storage

Legal requirements

The minimum amount of cold-water storage depends upon the Local Water Authority Regulations. They stipulate that the minimum actual capacities of cold-water cisterns are:

1. 114 litres if the cistern is used for a direct system where all the cold-water draw-offs are supplied directly from the main. The cistern is known as a 'feed' cistern because it feeds only the hot-water cylinder.
2. 227 litres if the cistern is used for an indirect system where only the drinking water is supplied directly from the main and all other fittings are supplied 'indirectly' from what is now termed a cold-water storage cistern.

Advantages of storage

The provision of a storage cistern gives the following advantages:

1. It provides a reserve of water which can reduce the effects of a mains failure.
2. It reduces the demand on the water main.
3. Pressure fluctuations on the main are reduced.
4. It reduces the pressure on the distributing pipes which supply WC, basins, baths, showers, etc.
5. Hot-water cylinders can be substantially lighter in construction than they would be if they had to resist the mains pressure.

6. Because a storage system reduces pressure fluctuations, thermostatic mixing valves for showers, etc., operate more efficiently.
7. The reduction of pressure reduces the risk of noise from water hammer and there is less wear on the valves.
8. The air gap between the ballvalve and the water level in the storage cistern prevents water being drawn back into the main, due to 'back siphonage'.

Disadvantages of storage

1. The installation cost is increased.
2. A storage cistern should be covered with a dust-proof lid, but this is frequently omitted and as a result the water can become dirty or even contaminated by fungus growths, or mice and birds.
3. Space, structural support, and insulation must be provided for the cistern.

Unvented hot water storage systems, now permitted by the Regulations, allow cold water direct from the main to feed the hot water storage cylinder and also to supply cold water draw-off taps.
Cold water storage inside the building is therefore not required.

Provision of storage

For larger buildings, the capacity of the cold-water storage cistern will have to be estimated, and Table 10.1 gives the storage requirements for various types of buildings to cover 24-hour interruption of supply.

Table 10.1 Provision of cold-water storage to cover 24-hour interruption of supply.

Type of building	Storage per person (litres)
Dwelling houses and flats	91
Hostels	91
Hotels	136
Offices without canteens	37
Offices with canteens	45
Restaurants — per meal	7
Day schools	27
Boarding schools	91
Nurses' homes and medical quarters	114

Example 10.1. *Calculate the capacity of the cold-water storage cisterns for a five-storey office block which will have a canteen. The population will be 80 persons on each floor and a 10-hour storage of water in case of interruption of supply has been decided.*

storage capacity for 24-hr interruption
$$\text{of supply} = 5 \times 80 \times 45 = 18\ 000 \text{ litres}$$

storage capacity for 10-hr interruption
$$\text{of supply} = \frac{18\ 000}{1} \times \frac{10}{24} = 7500 \text{ litres}$$

It is not always possible at the early design stage to know the number of people that will occupy a building, but the number and types of sanitary fittings will be known.

Table 10.2 may be used to find the storage requirements.

Table 10.2   Capacities of hot and cold water added together, required for single use of appliances

Appliance	Capacity in litres
Wash basin	
hand wash	5
hand and face wash	10
hair wash	20
Shower	40
Bath	110
WC	10
Washing machine	150
Sink	
wash-up	15
cleaning	10

Example 10.2. *Calculate the capacity of a cold-water storage cistern for a dwelling containing bath, WC, wash basin, and sink.*

During the peak demand period it is usual for a dwelling to allow consecutive use of a bath and single use of other fittings. The wash basin will normally be in the bathroom so this may be ignored; also the cold water to the sink will be supplied directly from the main so a smaller amount of cold water will be required to feed the hot-water cylinder.

$$
\begin{array}{rcl}
2 \text{ baths} \times 110 & = & 220 \\
1 \text{ WC} \times 10 & = & 10 \\
1 \text{ sink} \times 5 & = & \underline{5} \\
& & 235 \text{ litres}
\end{array}
$$

This is above the minimum capacity of 227 litres required by the regulations. If the mains supply is good, the water entering the cistern during the time that water is drawn off should permit a 227-litre storage being used satisfactorily.

The above method may also be used to estimate the minimum cold-water storage for larger buildings. The following steps may be taken:

1. Prepare a worksheet as shown in Table 10.3 and enter all types of sanitary appliances with their requirements obtained from Table 10.2.
2. Check the hourly pattern of use of sanitary appliances with the building owner and estimate the highest possible demand. For each hour of the day, multiply the number of times each type of appliance is likely to be used, by its capacity, and enter the product under the hour of day.
3. When all the sanitary fittings have been completed, add them under their respective columns and enter each total in the gross-demand row.
4. Hourly rates of water flow into the cistern should be obtained from the Local Water Authority and entered in the recovery-rate row. The figures

186

Table 10.3   Estimation of water consumption

Types of appliances and consumption (litres)	Hours of day																							
	1	2	3	4	5	6	7	8	9	10	11	12	13	14	15	16	17	18	19	20	21	22	23	24
Basins						30	50	100	100	50	50	100	50	50	50	50	150	100	100	50	100	100	100	50
Baths							550	550	550	330	330	330	330	330	220	220	330	550	550	550	220	220	220	220
Showers						80	320	320	200	160	160	160	120	200	200	80	80	320	320	320	160	160	80	80
WCs						100	200	200	100	100	50	50	200	50	100	50	50	200	200	200	200	200	200	100
Sinks						25	100	100	100	50	50	50	100	100	100	50	50	50	100	100	100	50	50	50
Gross Demand						235	1220	1270	1050	690	640	690	800	730	670	450	660	1220	1270	1220	780	730	650	450
Recovery rate						700	700	700	700	700	700	700	700	700	700	700	700	700	700	700	700	700	700	700
Net loss or gain						+ 465	− 520	− 575	− 350	+ 10	+ 60	+ 10	− 100	− 30	+ 30	+ 250	+ 40	− 520	− 570	− 520	− 80	− 30	+ 50	+ 250

in the gross-demand row should be subtracted to give net hourly loss or gain from the cistern.

5. The amount of water to be expected in the cistern at the end of each hour can then be plotted on the chart (see Fig. 10.1). The plotting should be commenced at the hour when the cistern is assumed to be full, i.e., early morning. Net the gains or losses incurred, in the hour when the water is being drawn off from the cistern, by adding or subtracting, and repeat this process until the graph is complete. If it shows that the cistern is likely to be nearly empty at any hour of the day, a larger cistern would be required and the graph redrawn.

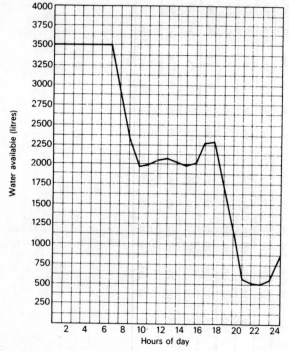

Fig. 10.1   Chart for storage capacity

Example 10.3 may help to clarify the use of the method.

Example 10.3.   *A three-storey hotel is to be constructed having the following sanitary fittings: 20 wash basins; 10 baths; 10 showers; 10 WCs; 5 sinks (cold water direct from the main).*

The water supply from the main is calculated to supply 700 litres per hour for refilling the cistern. It is proposed to install two 1750-litre storage cisterns interconnected so that one cistern may be used as a standby. Check if the storage proposed is satisfactory.

The estimated amounts of water used per hour starting from 06.00 hours and finishing at 24.00 hours are shown on Table 10.3.

From the net loss or gain, a graph can be plotted as shown in Fig. 10.1. It will be seen from the graph that about 500 litres of cold water will always be available in the cisterns so that a 3500-litre storage would be satisfactory.

If the graph showed that the cisterns were to be completely emptied at any time of the day, the storage allowed would have to be increased. It is advisable always to have the cisterns at least 5 per cent full during the worst period of the day and this addition will allow for the possibility of several draw-off points being used in rapid succession.

Hot-water storage and boiler power

The hot-water storage requirements for dwellings will vary according to the customs of the occupants, but a 136-litre storage capacity should be taken as the minimum.

In large buildings it is usually possible to make a reasonably close estimate. In most buildings there is a peak demand at least twice per day and ample storage must be provided to meet these demands.

In general, it is better to install a system with adequate storage rather than one with inadequate storage which would rely on rapid firing of the boiler at certain periods to meet the peak loads. Adequate storage will reduce the possibility of complaints and also enable the boiler to be fired steadily and economically. Greatly oversizing the hot-water storage, however, will increase the capital cost of the installation.

Table 10.4 may be used as a guide to the hot-water storage requirements for various types of buildings.

Table 10.4 Provision of hot-water storage .

Type of building	Storage per person (litres)	Type of building	Storage per person (litres)
Colleges and schools		Hospitals	
boarding	23	general	27
day	4.5	infectious	45.5
Dwelling houses	45.5	maternity	32
Factories	4.5	Nurses' homes	45.5
Flats	32	Hostels	32
Hotels (average)	35	Offices	4.5
		Sports pavilions	36

Example 10.4. *Calculate the hot-water storage requirements for a house having three occupants.*

storage required = 3 × 45.5 = 136.5 litres
a 136-litre capacity cylinder would be suitable

Table 10.5 gives the storage of hot water for restaurant and canteen kitchens.

189

Table 10.5 Hot-water storage requirements for restaurant and canteen kitchens for a 2-hour recovery period

Number of main meals served	Storage in litres
50	455
100	568
200	682
300	909
400	1137
500–600	1364
700–800	1818
900–1000	2273
1250	2841
1500	3410

Example 10.5. *Calculate the hot-water storage requirements for a factory having a total workforce of 500. The factory is to have a canteen kitchen which will prepare main meals for all the workforce.*

Table 10.4 shows that 4.5 litres per person of hot water will be required for ablution purposes and Table 10.5 shows that 1364 litres of hot water will be required for the canteen kitchen; therefore the total hot-water storage requirements will be:

$$ablution\ purposes = 500 \times 4.5 = 2250\ litres$$
$$canteen\ kitchen = 1364\ litres$$
$$total = 3614\ litres$$

If the population of the building is unknown, an estimate of the storage requirements may be made from the amount of hot water used by the various sanitary appliances.

Table 10.6 gives the amount of water used for various sanitary appliances.

Table 10.6 Capacity of hot water used at each appliance

Appliance	Capacity
Wash basin	
hand wash	1.5
wash	3
hair wash	6
Shower	13
Bath	70
Washing machine	70
Sink	
wash-up	15
cleaning	5

Example 10.6. *Estimate the hot-water storage requirements for a small hospital having the following sanitary appliances: 25 wash basins; 15 baths; 15 showers; 10 wash-up sinks.*

The amount of hot water stored must be sufficient to meet the peak demand periods and the boiler power sufficient to recover this heat in 2 hours. Assuming a 1-hour peak demand period, the hot-water storage may be estimated as follows:

25 basins used four times	$= 25 \times 4 \times 3 =$	300	
15 baths used twice	$= 15 \times 2 \times 70 =$	2100	
15 showers used three times	$= 15 \times 3 \times 13 =$	585	
10 wash-up sinks used three times	$= 10 \times 3 \times 15 =$	450	
		3435	

It would be advisable for a hospital to duplicate the storage calorifiers and boilers for standby purposes and each calorifier should contain two-thirds of the total storage, therefore each calorifier would have a capacity of:

$$\text{capacity of each calorifier} = 3435 \times \frac{2}{3} = 2290 \text{ litres}$$

Boiler power

The power in kW required for each boiler may be found from the following expression:

$$\text{boiler power} = \frac{\text{s.h.c.} \times \text{kg} \times \text{temperature rise (K)} \times 100}{\text{heating-up time in seconds} \times \text{efficiency}} \text{ kW}$$

where s.h.c. = specific heat capacity of water of 4.2 kJ/kg K

The following factors may be used:

(a) temperature rise from $10\,^\circ C$ to $60\,^\circ C = 50\,^\circ C$
(b) heating-up time or recovery period = 2 hours
(c) efficiency of plant = 70 per cent

$$\therefore \text{ boiler power} = \frac{4.2 \times 2290 \times 50 \times 100}{2 \times 3600 \times 70}$$

$$= 95.416$$

Two 96 kW boilers would be satisfactory.

Mixing quantities of water at different temperatures

It is sometimes necessary to calculate the resultant temperature of two quantities of hot and cold water when mixed.

The problem can be solved by the fundamental expression:

heat lost by hot water	= heat gained by cold water
heat lost by hot water	$= q_1 (t_h - t)$
heat gained by cold water	$= q_2 (t - t_c)$
where $\quad q_1$	= quantity of hot water in litres or m^3
t_h	= temperature of hot water in $^\circ C$
q_2	= quantity of cooler water in litres or m^3

$$t_c = \text{temperature of cooler water in } °C$$

$$t = \text{final temperature of mixture in } °C$$

By transposition

$$q_1 (t_h - t) = q_2 (t - t_c)$$

$$(q_1 \times t_h) - (q_1 \times t) = (q_2 \times t) - (q_2 \times t_c)$$

$$(q_1 \times t_h) + (q_2 \times t_c) = (q_1 \times t) + (q_2 \times t)$$

$$(q_1 \times t_h) + (q_2 \times t_c) = t(q_1 + q_2)$$

$$\therefore t = \frac{(q_1 \times t_h) + (q_2 \times t_c)}{q_1 + q_2}$$

Example 10.7. *A tank containing 50 litres of water at 70 °C has 20 litres of cold water at 10 °C mixed with it. Determine the resultant temperature of the whole. Ignore heat losses.*

$$t = \frac{(50 \times 70) + (20 \times 10)}{50 + 20}$$

$$t = 52.86 °C \text{ (approx.)}$$

Example 10.8. *A three-way thermostatic mixing valve allows 5 litres at 60 °C to mix with 8 litres at 40 °C. Determine the resultant temperature of the water discharging from the valve.*

$$t = \frac{(5 \times 60) + (8 \times 40)}{5 + 8}$$

$$= 47.69 °C$$

In ventilating and air-conditioning systems, recirculated air is often mixed with fresh air. The temperature of the resultant mixture may be found by use of the same equation.

Example 10.9. *In an air-conditioning system 15 kg of fresh air at 8 °C is mixed with 10 kg of recirculated air at 22 °C. Determine the resultant temperature of the mixture.*

$$t = \frac{(10 \times 22) + (15 \times 8)}{10 + 15}$$

$$= 13.6 °C$$

Example 10.10. *A recirculating fan hot-water convector heater is used to heat a room, the heat losses from which are 5 kW. If the convector handles 0.08 m³/s of air to provide a room air temperature of 22 °C, calculate the temperature of the air leaving the convector.*

heat capacity of air (HC) = 1.34 kJ/m³ K

$$kW = m^3/s \times HC \times (t_1 - t_2)$$

where
$$t_1 = \text{temperature of the air leaving the convector } (°C)$$
$$t_2 = \text{temperature of the air in the room } (°C)$$

By transposition

$$t_1 = \frac{kW}{m^3/s \times HC} + t_2$$

$$= \frac{5}{0.08 \times 1.34} + t_2$$

$$= 68.64\,^\circ C$$

$$= 69\,^\circ C \text{ (approx.)}$$

Expansion of materials

Various materials used in services do not expand equally when their temperatures
are raised. In order to determine the increase in length of a material due to an
increase of temperature, the coefficient of linear expansion of the material must
be known.

The coefficient of linear expansion is defined as the fraction of its original
length by which a substance expands per degree rise in temperature. Table 13.7
gives the coefficient of linear expansion of some common materials used in
services.

Table 10.7   Coefficient of linear expansion of common metals and plastics
used in services

Aluminium	
cast	0.000 025 5
sheet	0.000 023 0
Brass	
cast	0.000 018 7
sheet	0.000 019 3
Copper	
sheet	0.000 017 4
tube	0.000 016 9
Iron	
cast	0.000 010 2
Lead	
pipe and sheet	0.000 029 3
Steel	
mild	0.000 011 3
nickel chrome	0.000 015 3
invar	0.000 001 0
Tin	0.000 021 4
Zinc	
sheet	0.000 026 1
Polyethylene	
low density	0.000 28
high density	0.000 11
Polyvinyl chloride	
normal impact	0.000 05
high impact	0.000 08

Example 10.11. *Calculate the linear expansion of 15 m of copper tube when heated from 10°C to 70°C.*

expansion = coefficient of linear expansion × temperature rise × length

expansion = 0.000 016 9 × (70 − 60) × 15

= 0.015 21 m or 15.21 mm

Example 10.12. *A straight length of mild-steel heating pipe 200 m long will be heated from 15°C to 80°C. Expansion joints that will each accommodate 76 mm of the expansion are to be inserted in the straight length. Calculate the number of expansion joints that will be required.*

expansion = 0.000 011 3 × (80 − 15) × 200

= 0.1469 m or 146.9 mm

$$\text{number of joints} = \frac{146.9}{76} = 1.933$$

Therefore two expansion joints would be required.

Expansion of water

The expansion of water on being heated may be found from the following expression:

$$E = V \left[\frac{\rho_1 - \rho_2}{\rho_2} \right]$$

where E = expansion of water in litres or m³

V = volume of water before being heated in litres or m³

ρ_1 = density of water before being heated in litres or m³

ρ_2 = density of water after being heated in litres or m³

Table 10.8 gives the density of water at various temperatures.

Table 10.8 Density of water at various temperatures

Water temperature (°C)	Water density (kg/m³)	Water temperature (°C)	Water density (kg/m³)
0	999.80	60	983.00
4	1 000.00	63	982.00
10	999.70	65	980.70
15.6	990.00	66	980.10
20	998.00	70	977.50
25	996.50	71	977.00
30	995.00	74	975.00
38	993.00	80	972.00
40	992.00	82	970.40
46	989.00	91	965.00
50	987.50	100	958.00

Example 10.13. *A low-temperature hot-water heating system contains 12 m^3 of water at a temperature of 10 °C before being heated. Calculate the expansion of the water in the system when heated to a design temperature of 80 °C.*

from Table 10.8,

density of water at 10 °C = 999.70 kg/m^3

density of water at 80 °C = 972.00 kg/m^3

$$\text{expansion} = 12 \left[\frac{999.70 - 972}{972} \right]$$

$$= 12 \times \frac{27.7}{972}$$

$$= 0.342 \text{ m}^3 \text{ (approx.)}$$

Feed and expansion cistern (see Fig. 10.2)

Cover

Vent pipe

Cold water from main

40 mm
40 mm

Additional expansion space

Overflow pipe

Dip pipe to prevent cold-air entry into cistern

Normal expansion space

Depth of water about 100 mm above cold feed outlet

Extension for large cistern

Primary cold feed pipe

Fig. 10.2 Feed and expansion cistern

In a low-temperature heating system, a feed and expansion cistern is used to fill the system with cold water and also to accept the expanded water when the system heats up. In practice, the depth of cold water in the cistern is about 100 mm and the rest of the space is left for the expanded water, plus about $33\frac{1}{3}$ per cent for additional expansion for exceptionally cold weather when the boiler may be operated at a higher temperature.

The gas laws

Boyle's law (see Fig. 10.3)

This law states that for a fixed mass of gas at constant temperature, the volume is inversely proportional to the pressure:

$$P = \frac{C}{V}$$

or $PV = C$

where

P = absolute pressure

V = volume

C = constant

Also $\quad P_1 V_1 = P_2 V_2$

The application of the above equation enables any difference in volume to be determined.

If the volume of the gas in the cylinder is halved the pressure is doubled and vice versa also $3V = \frac{1}{3}P$ and $\frac{1}{3}V = 3P$, etc

Fig. 10.3   Boyle's law

Example 10.14 (see Fig. 10.4).   *The initial volume of air at atmospheric pressure in a pneumatic cylinder for boosting cold water is 3.5 m³. When the pumps are switched on this volume is reduced to 0.8 m³. Calculate the pressure of air in the cylinder for boosting the water. Atmospheric pressure can be taken as 101.33 kPa.*

$$P_1 = 101.33 \text{ kPa}$$

$$V_1 = 3.5 \text{ m}^3$$

$$V_2 = 0.8 \text{ m}^3$$

$$P_1 V_1 = P_2 V_2$$

$$\therefore P_2 = \frac{P_1 V_1}{V_2}$$

$$= \frac{101.33 \times 3.5}{0.8}$$

196

Fig. 10.4 Automatic pneumatic cylinder

$$P_2 = 441.875 \text{ kPa (absolute)}$$
$$= 441.875 - 101.33 \text{ (gauge)}$$
$$= 340.545 \text{ kPa (gauge)}$$

Theoretically, this would force the water up to a height of:

$$\text{height} = \frac{340.545}{9.81} = 34.7 \text{ m (approx.)}$$

This would result in a height of about 30 m, depending upon the frictional losses due to the pipes and fittings.

Charles' law

This law states that for a fixed mass of a perfect gas at constant pressure, the volume is directly proportional to its absolute temperature. The law is derived from the fact that a given mass of a perfect gas will change its volume 1/273 of its volume at $0\,^\circ$C for a change in temperature of $1\,^\circ$C, provided that the pressure remains constant.

By combining Boyle's law and Charles' law, a formula is obtained which takes into account the effect of varying pressure, volume, and temperature (see Fig. 10.5):

$$\frac{P_1 V_1}{T_1} = \frac{P_2 V_2}{T_2}$$

where
P_1 = initial pressure

P_2 = new pressure

V_1 = initial volume

V_2 = new volume

T_1 = initial temperature

T_2 = new temperature

Fixed mass of gas with a change in temperature

Piston

Cylinder

Piston

If the volume remains constant
the pressure of gas will change proportionally with absolute temperature

Fig. 10.5   Combination of Boyle's and Charles' laws

Example 10.15.   *A compressed air cylinder has a volume of 5 m³ and contains air at atmospheric pressure and a temperature of 20 °C. If the air temperature in the cylinder is raised to 80 °C, calculate the new pressure in the cylinder.*

$$P_1 = 101.33 \text{ kPa (absolute)}$$
$$V_2 = 5 \text{ m}^3$$
$$V_1 = 5 \text{ m}^3$$
$$T_2 = 70 + 273 = 343 \text{ K}$$
$$T_1 = 20 + 273 = 293 \text{ K}$$

$$\frac{P_1 \, V_1}{T_1} = \frac{P_2 \, V_2}{T_2}$$

$$P_2 = \frac{P_1 \times V_1 \times T_2}{V_2 \times T_1}$$

$$= \frac{101.33 \times 5 \times 343}{5 \times 293}$$

$$= 118.62 \text{ kPa (absolute)}$$

$$= 118.62 - 101.33 = 16.29 \text{ kPa (gauge)}$$

Drainage testing

When drainage pipelines are tested by use of air, it is essential to take into account the fall in air pressure due to the air temperature. If the air that is pumped into the pipeline is higher in temperature than the air already inside the pipeline, its volume will be reduced, showing a drop in the U-gauge water level. In order to obtain a correct reading of the U-gauge, about 15 minutes should be allowed so that the air pressure stabilises.

Example 10.16 (see Fig. 10.6). *A 225 mm soil stack is subjected to an air test recorded by a water filled U-gauge. At an initial temperature of 20 °C the gauge showed a reading of 40 mm head of water. Calculate the head of water in the gauge when the air temperature inside the pipe falls to 15 °C.*

Fig. 10.6 Air test on soil stack

$$\frac{P_1 V_1}{T_1} = \frac{P_2 V_2}{T_2}$$

$$\frac{40 \text{ mm} \times V_1}{20 + 273} = \frac{\text{head of water} \times V_2}{15 + 273}$$

v_1 and V_2 are equal and are therefore cancelled

$$\therefore \frac{40}{293} = \frac{\text{head of water}}{288}$$

199

$$\text{head of water} = \frac{40 \times 288}{293}$$

$$= 39.32 \text{ mm (approx.)}$$

Example 10.17. *5 m^3 of a gas at an initial pressure of 50 kPa (gauge) and at a temperature of 30 °C is compressed in a cylinder to 3 m^3, the pressure of the gas being raised to 155 kPa (gauge). Calculate the rise in temperature of the gas after compression.*

$$P_1 = 50 + 101.33 \text{ kPa (absolute)}$$

$$P_2 = 155 + 101.33 \text{ kPa (absolute)}$$

$$V_1 = 5 \text{ m}^3$$

$$V_2 = 3 \text{ m}^3$$

$$T_1 = 30 + 273 = 303 \text{ K}$$

$$\frac{P_1 V_1}{T_1} = \frac{P_2 V_2}{T_2}$$

$$T_2 = \frac{T_1 \times P_2 \times V_2}{P_1 \times V_1}$$

$$= \frac{303 \times 256.33 \times 3}{151.33 \times 5}$$

$$= 308 \text{ K (approx)}$$

$$\text{rise in temperature} = 308 - 303$$

$$= 5°C$$

Example 10.18. *16 m^3 of air at 10 °C and at a pressure of 180 Pa (gauge) is passed through a heater battery. If the temperature of the air is raised to 50 °C and the pressure falls to 80 Pa (gauge), calculate the change in volume of the air.*

$$P_1 = 101.33 + 0.180$$

$$P_2 = 101.33 + 0.08$$

$$V_1 = 16 \text{ m}^3$$

$$T_2 = 50 + 273 = 323 \text{ K}$$

$$T_1 = 10 + 273 = 283 \text{ K}$$

$$\frac{P_1 V_1}{T_1} = \frac{P_2 V_2}{T_2}$$

$$V_2 = \frac{P_1 \times V_1 \times T_2}{T_1 \times P_2}$$

$$= \frac{101.51 \times 16 \times 323}{283 \times 101.41}$$

$$= 18.279 \text{ m}^3$$

$$\text{change in volume} = 18.279 - 16$$

$$\fallingdotseq 2.279 \text{ m}^3$$

Questions

1. State the Water Authorities' legal requirements for the minimum storage of cold water in buildings and explain the merits and demerits of cold-water storage.

2. Calculate the capacity of the cold-water storage cisterns required for a hotel having 200 residents, including staff. A 24-hour storage is required in case of interruption of supply.

Answer: 27 200 litres

3. A five-storey office block is to be constructed having the following sanitary fittings on each floor: 6 WCs, 8 wash basins, and 1 sink. Prepare a work-sheet of the probable amount of water used during the time between 07.00 and 18.00 hours and from the values draw a graph to find the minimum cold-water storage requirements. The water supply from the main is calcu-lated to supply 9000 litres per hour for refilling the cistern.

4. Calculate the hot-water storage requirements for a general hospital having 600 patients and staff.

Answer: 16 200 litres

5. If the hot-water calorifiers for the hospital in question 4 are to be duplicated for maintenance or repair, calculate the power in kW of each boiler when each calorifier will have a capacity of $\frac{2}{3}$ of the total storage. Use the following factors:

   (a) Temperature rise of water = $55\,^{\circ}$C
   (b) heating-up time or recovery period = 2 hours
   (c) efficiency of plant = 60 per cent
   (d) specific heat capacity of water = 4.2 kJ/kg K

Answer: 57.75 kW (or 58 kW approx.)

6. Estimate the hot-water storage for a sports pavilion having the following sanitary fittings: 10 showers, 8 WCs, 8 wash basins, and 2 cleaning sinks.

7. A tank containing 150 litres of cold water at 10 $^{\circ}$C has 100 litres of hot water at 80 $^{\circ}$C mixed with it. Determine the resultant temperature of the mixture. Ignore heat losses.

Answer: 38 $^{\circ}$C

8. Calculate the linear expansion of 60 m of steel tube when heated from 20 $^{\circ}$C to 80 $^{\circ}$C.

Answer: 46.8 mm

9. A low-temperature hot-water heating system contains 45 m^3 of water at a temperature of 15.6 $^{\circ}$C before being heated. Calculate the expansion of the water in the system when heated to a design temperature of 74 $^{\circ}$C.

Answer: 0.6923 m^3

10. Define Boyle's law and state how it may be applied to various problems in services when the temperature remains constant.

Appendices

SI units (International System of units)

In 1971, the Council of Ministers of the European Community (EEC) decided to commit all members countries to amend their legislation in terms of SI units. The United Kingdom had already decided that SI units would become the primary system of measurement and legislation is established in several countries.

There are three classes of SI units, namely:

1. Base units
2. Supplementary units
3. Derived units

Base units

The SI system is based on seven units.

Quantity	Unit	Symbol
Length	metre	m
Mass	kilogram	kg
Time	second	s
Electric current	ampere	A
Thermodynamic temperature	kelvin	K
Luminous intensity	candela	cd
Amount of substance	mole	mol

Note: For ordinary temperature and the difference between two temperatures, i.e., temperature interval, the degree Celsius ($^\circ$C) is used.

Supplementary units

Quantity	Unit	Symbol
Plane angle	radian	rad
Solid angle	steradian	sr

The radian is the angle between two radii of a circle which cut off on the circumference an arc equal in length to the radius.

The steradian is an angle which, having its vertex in the centre of a sphere, cuts off an area of the surface of the sphere equal to that of a square having sides of length equal to the radius of the sphere.

Derived units

These are expressed algebraically in terms of base units and/or supplementary units. The main units are:

Quantity	Name of derived units	Symbol	Units involved
Frequency	hertz	Hz	$1\ Hz = 1\ s^{-1}$ (1 cycle per second)
Force	newton	N	$1\ N = 1\ kg\ m/s^2$
Pressure and stress	pascal	Pa	$1\ Pa = 1\ N/m^2$
Work, energy and quantity of heat	joule	J	$1\ J = 1\ N\ m$
Power	watt	W	$1\ W = 1\ J/s$
Quantity of electricity	coulomb	C	$1\ C = 1\ A\ s$
Electrical potential, potential difference, electromotive force	volt	V	$1\ V = 1\ W/A$
Electric capacitance	farad	F	$1\ F = 1\ A\ s/V$
Electric resistance	ohm	Ω	$1\ \Omega = 1\ V/A$
Inductance	henry	H	$1\ H = 1\ V\ s/A$
Luminous flux	lumen	lm	$1\ lm = 1\ cd\ sr$
Illuminance	lux	lx	$1\ lx = 1\ lm/m^2$

Multiples and sub-multiples of SI units

Factor	Prefix		
	Name		Symbol
One billion (one million million)	10^{12}	tera	T
One thousand million	10^9	giga	G
One million	10^6	mega	M
One thousand	10^3	kilo	k
One hundred	10^2	hecto	h
Ten	10^1	deca	da
One tenth	10^{-1}	deci	d
One hundredth	10^{-2}	centi	c
One thousandth	10^{-3}	milli	m
One millionth	10^{-6}	micro	μ
One thousand millionth	10^{-9}	nano	n
One million millionth	10^{-12}	pico	p

Units for general use

Quantity	Unit	Symbol	Definition
Time	minute	min	1 min = 60 s
	hour	h	1 h = 60 min
	day	d	1 day = 24 h
Plane angle	degree	°	$1° = (\pi/180)$ rad
	minute	′	$1′ = (1/60)°$
	second	″	$1″ = (1/60)′$
Volume	litre	l	$1\ l = 1\ dm^3$
			$(1000\ l = 1\ m^3)$
Mass	tonne	t	$1\ t = 10^3$ kg

Surface areas, volumes and perimeters

Surface areas

Triangle

$$\text{Area} = \frac{bh}{2}$$

or

$$\text{Area} = \sqrt{s(s-a)\ (s-b)\ (s-c)}$$

Where $s = \dfrac{a + b + c}{2}$

Rectangle

$$\text{Area} = lb$$

Parallelogram

$$\text{Area} = bh$$

Trapezium

$$\text{Area} = \frac{b + a}{2} \, h$$

Irregular polygons

Area = area *abe* + area *bce* + area *cde*

Rhombus

$$\text{Area} = \frac{ab}{2}$$

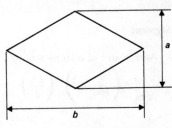

Circle

$$\text{Area} = \pi r^2 \quad \pi = 3.1416 \text{ (usually taken as 3.142)}$$

or

$$\text{Area} = \frac{\pi}{4} \, d^2$$

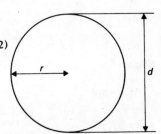

Annulus

$$\text{Area} = \pi(R^2 - r^2)$$

or

$$\text{Area} = \pi(R - r)(R + r)$$

Parts of a circle

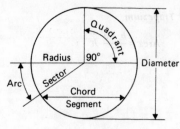

Sector

$$\text{Area} = \frac{\theta}{360}\, \pi r^2$$

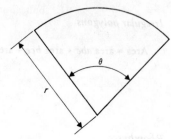

Segment

Area = area of sector – area of triangle

$$= \left(\frac{\theta}{360}\, \pi r^2\right) - \left(\frac{bh}{2}\right)$$

Ellipse

$$\text{Area} = \pi ab$$

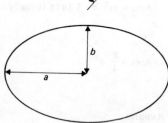

Volumes

Pyramid

$$\text{Volume} = \frac{\text{Area of base} \times h}{3}$$

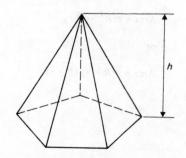

206

Cone

$$\text{Volume} = \frac{\text{Area of base} \times h}{3}$$

$$= \frac{\pi r^2 h}{3}$$

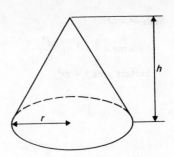

Rectangular prism

Volume = length × breadth × depth
= *lbd*

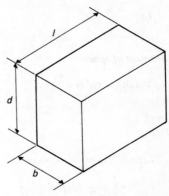

Circular prism or cylinder

Volume = $\pi r^2 h$
or
Volume = $\frac{\pi d^2}{4} h$

Sphere

Volume = $\dfrac{4}{3}\pi r^3$ or $\dfrac{\pi d^3}{6}$

Surface area = $4\pi r^2$

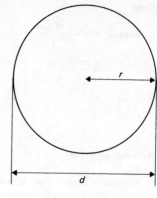

Segment of sphere

Volume = $\pi h^2 \left(r - \dfrac{1}{3}h\right)$

Circular ring

Volume = $2\pi^2 r^2 R$

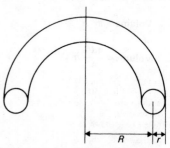

Perimeters

Square = 4 × length of one side
Rectangle = (2 × length) + (2 × breadth)
Triangle = sum of lengths of each side
Circle = πD or $2\pi r$

Arc of circle = $\dfrac{\theta}{360}$ × $2\pi r$ where θ is the angle subtended at centre of circle

Ellipse = $\pi(a + b)$ where a and b are half the lengths of the long and short axes

Index

209